About Island Press

Island Press is the only nonprofit organization in the United States whose principal purpose is the publication of books on environmental issues and natural resource management. We provide solutions-oriented information to professionals, public officials, business and community leaders, and concerned citizens who are shaping responses to environmental problems.

In 2000, Island Press celebrates its sixteenth anniversary as the leading provider of timely and practical books that take a multidisciplinary approach to critical environmental concerns. Our growing list of titles reflects our commitment to bringing the best of an expanding body of literature to the environmental community throughout North America and the world.

Support for Island Press is provided by The Jenifer Altman Foundation, The Bullitt Foundation, The Mary Flagler Cary Charitable Trust, The Nathan Cummings Foundation, The Geraldine R. Dodge Foundation, The Charles Engelhard Foundation, The Ford Foundation, The German Marshall Fund of the United States, The George Gund Foundation, The Vira I. Heinz Endowment, The William and Flora Hewlett Foundation, The W. Alton Jones Foundation, The John D. and Catherine T. MacArthur Foundation, The Andrew W. Mellon Foundation, The Charles Stewart Mott Foundation, The Curtis and Edith Munson Foundation, The National Fish and Wildlife Foundation, The New-Land Foundation, The Oak Foundation, The Overbrook Foundation, The David and Lucile Packard Foundation, The Pew Charitable Trusts, The Rockefeller Brothers Fund, Rockefeller Financial Services, The Winslow Foundation, and individual donors.

About Redefining Progress

Since 1994, Redefining Progress has criticized the use of economic growth indicators as measures of progress, called for the development of indicators that tell us how people, the environment, and the economy are doing, and advanced market-based incentives to enhance the well-being of people and the planet. A nonprofit, nonpartisan policy research center, Redefining Progress furthers its four core ideas: exposing the inadequacy of equating economic growth with progress; facing and embracing nature's limits as a path to true sustainability; promoting pricing systems that incorporate the social and environmental costs of products; and recognizing and building the natural and social value of common assets. Combining critique with recommendations for positive action, Redefining Progress works with opinion leaders and the general public to infuse these issues into public discourse and public policy. Redefining Progress is supported by foundations and individuals. Partial support for this book was provided by The Ford Foundation.

TRIUMPH
of the
MUNDANE

TRIUMPH
of the
MUNDANE

THE UNSEEN TRENDS THAT SHAPE OUR LIVES AND ENVIRONMENT

HAL KANE

ISLAND PRESS
Washington, D.C. ♦ Covelo, California

Library of Congress Cataloging-in-Publication Data

Kane, Hal.
  Triumph of the mundane : the unseen trends that shape our lives and environment / Hal Kane.
      p. cm.
Includes bibliographical references and index.
  ISBN 1-55963-715-3
  1. Environmental degradation. 2. Nature—Effect of human beings on.
3. Human ecology. I. Title.
  GE140 .K36 2001
  304.2'8—dc21
                              00-011702

Printed on recycled, acid-free paper

Printed in Canada
10 9 8 7 6 5 4 3 2 1

To Noah Kane Melnick

Contents

Acknowledgments

As always, more people deserve thanks than ever can be thanked. First, however, deep appreciation goes to the Ford Foundation, whose grant financed the writing of this book as well as the research and development of some of its ideas. The Ford Foundation, as everyone knows, has a long and impressive history of making such grants, and its support of Redefining Progress and organizations like it has had remarkable value.

The individuals who have reviewed draft copies of this manuscript, or who have contributed substantially to it in other ways, include Jeff Rivkin, Mark Valentine, Craig Rixford, Cliff Cobb, Durwood Zaelke, Alan AtKisson, Norman Myers, Jonathan Rowe, Peter Barnes, and Terrence McNally. Their time and insight are appreciated greatly. Without such colleagues, writing is not only more difficult but also much less rewarding.

Special thanks go to Todd Baldwin, my editor at Island Press, who has contributed enormously to this manuscript. And, of course, thanks go to all of the people at Island Press who have done the crucial work of promotion, layout and design, and all aspects of its production. Without their energy and work, there could be no book.

Several other institutions have been important to this effort as well. Part of the manuscript was written at the Mesa Refuge, a writers' retreat in Pt. Reyes Station, California, dedicated to the exploration of the connections between ourselves, nature, and our economy and society.

All of the people at the Worldwatch Institute deserve thanks for their years of work to develop the thinking and the information available about the world's environmental, social, and economic issues. The idea behind this book began to be formed in my mind during the five years I spent at

Worldwatch, in Washington, D.C., researching and writing about economics, hunger, the environment, and refugees.

Redefining Progress, which recently moved from San Francisco to a new home in Oakland, California, has provided a home for thinking about topics like the speed of society, the inadequacy of the GDP as a measure of well-being, and other fundamental but alternative topics of public policy. These issues deserve much greater attention from the mainstream media and our public institutions, and we are fortunate to have an organization like Redefining Progress to try to broaden the attention received by such topics. Its sponsorship of this book is, of course, greatly appreciated.

Introduction

The Physical and Private Changes
That Have Shaped America

At 224 years of age, the United States is one of the oldest countries in the world. It is older than Italy, which was a group of independent city-states whose peoples spoke different dialects as recently as 120 years ago. It is more enduring than India, which, despite an ancient language and culture, created its present form of government and its present shape only during the last fifty years of the twentieth century. The United States has the longest enduring written constitution in the world. The institution of the presidency has remained since George Washington held that position.

Amid this remarkable constancy, how can it be that on the world stage the United States is notorious for change?

Our legislative and judicial systems have certainly evolved over time, but they are clear and direct descendents of the original system set up in the 1700s. Many of the issues that they confront have remained: racism endures, the national defense continues to be a preoccupation, joblessness is, as always, a national fear. Taxes and their avoidance remain a dominant topic of public discourse, and a cause of private resentment. The country's borders have changed little since the Civil War restored the Union and William Seward negotiated the purchase of Alaska, securing for the United States the westernmost reaches of the continent. Even in dress, American style has remained fairly constant—Americans have always worn European business suits and dresses, even if the lapels have widened and narrowed over time.

So what gives the United States its reputation for change?

The thesis of this book is that this reputation for change is a result of different phenomena, ones that are less about government and language,

1

borders and national defense than our textbooks suggest. These phenomena constitute a historical pattern that has been absent from many history books and our national dialogue—a set of changes that have shaped our country, often without involving political debate. This other history is physical. It is about change in the devices that make people's lives easier, in the amount of space that people have to live in, in how much distance they travel in a day, in how often they move to new houses, in the landscape of the country. It is about change that is not based on beliefs or rights or political ferment.

The laws driving these trends are not written by Congress, but rather are the laws of physics and chemistry and biology that determine how fast people can travel and what they can build and how healthy they can be. One of their common themes has to do with freedom—but not in the classic sense of freedom from government abuses. Today, distances that could long only be traveled in months or years are wiped away in a few hours by an airplane. This has freed people from the constraints of location. Meals that were grown, harvested, cooked, and served with great labor are now obtained with a moment's notice. This has freed people from the farm and from some housework. Today, tasks that required slow, laborious devotion are done quickly by assembly lines or computers. Clothes that were sewn or knitted are, today, bought—and women and men are freed to put their efforts elsewhere, as well as to have more clothes. Political freedom was critical to the origins of the United States, and remains critical. But as time has gone on, new, nonpolitical freedoms have taken center stage.

People's obligations to their families, which once kept them home instead of out exploring the world, have given way as children move out of the house at young ages and often live across the country from their parents. Duty to children, parents, or grandparents can be met sometimes by sending money in an envelope instead of through time spent at home. This partial freedom from family obligations has changed every aspect of our experience. No law passed by Congress can match the fundamental nature of such change; few decisions by an American court can affect people as deeply. The changing connections to the family are among the most important components of an unsung history of the United States.

Our lives are accelerating: the number of miles Americans travel in a day, the number of calculations we expect from our computers in a second, the products of our assembly lines, our expectations for what a person accomplishes in a day. The number of people living alone has risen dramatically as well, and changed our relationships with other people, our expectations for privacy, and our responsibilities. Our connections to the places where we live and where we are born are changing, just as our connections to the family are changing, and our connections to tradition, to groups of people, to nature, between rich and poor, and to other parts of the country are all evolving. The number of possessions that we own has grown dramatically over the years as well, possibly to compensate for gaps in our networks of human relationships, or possibly borne of the marriage of acquisitiveness and affluence.

These unheralded trends have slowly ascended to a position of governance in our lives today, overtaking the traditional roles of religious teaching, political discourse, and even family dinners. The country's official history is shaped by 535 senators and members of Congress who work part of the year to write new laws, often according to longstanding traditions— and by small numbers of other individuals. But across the country, more than one hundred million people work every day to make more and better consumer goods and provide services to customers. In Silicon Valley, about half a million people work to double the speed of information every eighteen months.[1] In every state, construction workers bulldoze hillsides for new suburbs and put up hospitals for better health care. Their combined products overwhelm the staples of college history tomes: judicial precedents and electoral formations.

All of these changes, from the building of roads to the burning of fossil fuels to people's choices at the shopping center and the affordability of living alone, have a direct and dramatic effect on the natural environment. Meanwhile, it is often the case that legal precedents and decisions by courts do not take the environment into account; they concentrate instead on property rights, contractual obligations, and economics. Even legislation directed at environmental issues, like the Clean Air Act, may have less effect on nature than the actions of Silicon Valley's use of materials and energy to build computers, or Detroit's use of the same to build cars. When it comes to the environment, the underlying trends recounted here are the primary events.

Environmental problems are chronicled in the media and in books of statistics as polluted air and water, shrinking forests, eroded soil, unstable climate, and so forth. But many of those changes are the physical manifestations of the speeds at which people travel, their quests for privacy and more living space, their expectations for material possessions, and other aspects of contemporary lifestyle. At their roots, environmental problems are not only pollution or waste but can be seen as the physical manifestation of our desires for mobility, privacy, comfort, disposability of goods, and other common goals.

If we treat environmental issues as distinct from lifestyle choices, we separate our actions from their environmental consequences, and we also let some of the fundamental currents of our history go undiagnosed. We should see these environmental trends as part of the evolution of the goals and priorities emerging from our country's history. We can read today's environmental issues, in part, as the embodiment of our social values, priorities, and doctrines.

The chapters of this book begin to delineate a powerful set of trends that are shaping the United States, that consist of living patterns and shopping sprees more than legislative processes or wars. It is a story of how fast people work and travel, and who they live with. No treaties or laws are discussed, but motels and drive-in movie theaters do receive attention, as do loneliness and isolation. Viewed individually, these things may seem beyond the scope of public affairs. Taken together, however, they can be seen as the basic elements of the story of our times. Yet the relatively small amount of research that has been done on these lifestyle issues mostly resides in obscure parts of government agencies like the Census Bureau and the Transportation Department, sometimes in academia, and occasionally in articles in demographic magazines read by advertising executives. The changes that have swept our personal lives deserve a much larger reception than can be found in the library of the Commerce Department or from the readership of a few journals. They are central to what happens to Americans, to our economy, and to our politics.

The Ascendancy of the Mundane

Today's Americans are among the first people in history to enjoy inexpensive antibiotics and prescription drugs, enticing electronics, and walk-in closets. People's loyalty to, and use of, new comfort-generating products like pharmaceuticals exists somewhat in proportion to the number of sneezes they have escaped during allergy season and the number of headaches cured by aspirin.

Such concrete innovations shape consciousness, and maybe even usurp part of the passion of public life. They are part of the ascendancy of practical little objects, like over-the-counter pills and walking shoes, that people can pick up with their hands, over principles and beliefs. There are others. Plumbing has saved lives by reducing disease, and has brought warm, daily showers. Better mattresses may help people have a better night's sleep. These are contributions to our well-being that are made by pharmaceutical companies, engineers, and technicians, and they may mean more to most Americans than the contributions made by their elected officials or even by their neighbors and community members. The success of providing some relief from allergies or pain contributes directly to people's comfort and their ability to function. No politicians can boast so much success in addressing their constituents' needs!

America is the most successful country ever at providing comfort and convenience to its people, and many of us are responding by placing our confidence in merchandise. All Americans use the objects made for us by engineers. Meanwhile, less than 5 percent of adult Americans engage in any kind of political activity aside from voting.[2] Since the 1960s, memberships have fallen dramatically in parent–teacher associations; women's organizations such as the League of Women Voters; the Boy Scouts, whose membership is down 26 percent since 1970; the Red Cross, which is down 61 percent since 1970; and the Lions Club, Elks, Shriners, Jaycees, and Masons. (For more data on political disengagement, see Chapter 2.) Meanwhile, the sales of myriad kinds of merchandise are booming.[3]

The greater share of the merchandise for sale in our stores and on our dotcom sites has been developed relatively recently in history, much of it since World War II.[4] People look to the merchandise not only for relief from their headaches, but also for solutions to deeper problems, like loneliness or fear. More self-help books sell every year than books about participatory politics, by far.[5] This is because self-help books claim to address deeper needs than do politics. Car dealers sell mobility and pharmacies sell relief. But politics and moral teachings offer abstractions that many people do not accept as valuable.

People have many choices today. We can buy food, cooked or uncooked, twenty-four hours a day. We can make phone calls from airplanes. We can participate in religious organizations or not, as we choose. But much remains for us to desire. People now seek freedom from boredom, and try to buy that freedom with devices: televisions, stereos, vacations, restaurant dinners. We try to free ourselves from unwanted jobs by using more efficient computers and faster assembly lines. But freedom evades many of those attempts, much like a fast car that gets nowhere when stuck in traffic on the way to work.

The search for new freedoms and goals is the tension of our time. The ability to leave the office. The opportunity to go off by oneself. It is privacy, space, speed, anonymity, and separation from whatever one finds undesirable that we seek in place of justice and liberty. The organizing principles of our time no longer center on political freedom but instead include the pursuit of convenience and comfort and health.

Many people complain that American politics is shallow and manipulated by special interests, and that true goals like justice and integrity are lost in our politics. But this assumes that justice and integrity *are* our true goals, rather than possession of objects and consumption of conveniences. Rather than being met by justice today, many of our goals are being met (i.e., bought) successfully, from medical care to comfortable shoes to machines that save us from hazardous work. It could be that our public discourse is stronger than it seems. Maybe it is a physical dialogue. Automobile advertisements that show a mother and daughter saved from an accident by antilock brakes may speak more directly and loudly to many people today than political advertisements showing candidates who claim to have a strong record on schools. The marketplace hosts an energetic banter, covering topics from safety to opportunity to planning for retirement to quality exercise. This banter holds the attention of many people who do not vote and do not read a newspaper.

In our music and art, the physical and banal daily details are taking hold. Rock songs from R.E.M. to Lou Reed list toy Tonka trucks, alleyways, and bars.[6] Rap music is filled with references to television sitcoms and brand names of clothing. Jerry Seinfeld's show is filled with kitchen gadgets that don't work right, catsup bottles in the diner, antics in the car, and the details of city life. Andy Warhol was not the last painter to portray household items—indeed, his ideas have invaded the works of other artists as well, an enduring triumph of the mundane.

Today's art and music are addressing the concerns on the minds of many people. They, more than our political dialogues, are reflecting the realities that are evolving in American culture today, such as whether or not we can find a parking space or how short of time we all feel. It is no accident that those parts of our political discourse, like zoning codes or gas prices, that do affect our physical surroundings are also the most contentious. Even discussions of "family values" often fail to include the things that most affect family life—such as how often families move to a new home or new state, or how far they have to drive to get to a good school or a job. In a recent state election in New Jersey, many surveys reported that the most important issue in the campaign was actually the price of automobile insurance. Those organizations that can speak to practical issues and offer the devices needed to deal better with daily pressures will win legislative debates and lead the country.

America could have a lively discussion about these topics. It could update our "national speed" of miles traveled every day in the newspaper alongside the Dow Jones Industrial Index. Our television stations and magazines could account for how many people live alone, and measure it the way they count the "Index of Leading Economic Indicators." Yet many of these changes have gone almost unchronicled. Data on how many miles Americans travel in a day, on average, have never been published. Data on how many people live alone are available only in an unnoticed brief from the Census Bureau. Statistics on how often American families move to new homes and new states are available from a monograph published by the Population Reference Bureau and have gotten only a small amount of attention in the popular press. Surprisingly, even though many people have a sense that ownership of appliances is rising, most data on ownership are owned, and rarely published, by industry associations.

The same could be said for data and information about many other topics that concern what is happening to the American people. This book draws together some of the most important physical and social changes that are shaping our future, but which have not gotten the attention that they deserve. Taken together, they draw a portrait of an evolving national character.

A New National Character

Alfred North Whitehead once said that people think in abstractions but live in detail.[7] The founding principles of the country are historical abstractions to many people. Americans are living in myriad details of decorated athletic shoes and fancy coffee drinks and kitchen appliances that come in all shapes and sizes. It is for these goals that many Americans work all day. Not to guarantee their right to speak or to worship but rather to secure their ability to drive a nice car or pay for an airplane ride and hotel room while on vacation.

The American dream has long been a home in the suburbs on a little piece of land, with a car in the driveway, and perhaps even a white picket fence. This dream is one of ownership rather than one of political freedom and justice for all. If we believed that freedom were threatened, then maybe we would rise to the occasion with the passion and commitment of our ancestors. But most of us do not believe that freedom or liberty is threatened. When we measure how the country is doing, it is not justice we measure, but gross national product, a quantity of material production and services.

Not all Americans share this unprecedented material abundance. For those of us who don't, the old American ideals may speak loudly. From such affronts as police brutality to illegally low wages and work conditions to toxic waste dumps located in poor communities and urban ghettos, many people may never have felt so strongly the need for the classic trio of "life, liberty, and the pursuit of happiness."

For other people, though, this moment in history is one of great success. It is the place where the founding beliefs of the country were intended to take us: a place where many Americans live free of poverty and anger, take basic human rights for granted, and have moved on to a concern for enjoying the benefits of those assets. The majority of Americans now alive have always had freedom of expression and freedom of religion, as well as refrigerators and freeways.

There is a danger of a split between the people who are benefiting from this moment in history and those who are not, between rich and poor, or those with opportunities and those without. Increases in the speed of transportation and the power of computers are available only to those who can afford them. Likewise, a country that caters to those who live alone will not provide for those whose families cannot even afford enough space for dignity. The host of possessions that affect the lives of many Americans do not exist in the homes of many other Americans. This disconnection between the people who can take advantage of today's rising speeds and abundances and those people who cannot is the most threatening aspect of our rapidly developing physical history.

In the last thirty years, many commentators have lamented that the United States has not made overwhelming progress in civil rights. But it has expanded living space per person. It has further spread the ownership of such objects as videocassette recorders and cars, but the gap between rich and poor has grown. A look into the differences between the areas in which we have failed and those in which we have succeeded offers a glimpse into the essence of what Americans are really working toward. It may be a glimpse into what Americans really care about. The reason why more progress has been made with possessions and living space than with civil rights is that America is putting more of its energy, work, and sweat toward those commodities than toward building a just and equitable society. We are building new machines more effectively than we are working toward abstract American ideals like equality and justice.

What if the workforce that gets up every morning and goes to work in San Jose, California, to build Internet search engines and Web site animation software could be redirected toward building equal rights and equal opportunities for all people? What if the energy and funding behind the U.S. military were behind environmental restoration or education for inner-city children or immunizations for those who lack them?

These questions are naive. The energy and work put toward marketable products cannot be redirected toward goals that must come instead from public discourse and a sense of common purpose. Work makes products. But a stronger, more cohesive society comes from shared values, not assembly lines.

But it is even more than that. Our successes at material comfort anesthetize us from the symptoms of public failure. When we shift our efforts toward the production of physical objects instead of toward responsibility for other people, then we focus less on those other people, and we insulate ourselves from their problems. When we replace the values that we share with other people by substituting for them with new merchandise, then we take away from our interest in assuring the well-being of our neighbors. As we succeed in surrounding ourselves with the objects of success we separate ourselves from the people who are not succeeding.

To many in the middle and upper classes, the people who dominate in the economy, the threats of the day are cholesterol and heart disease, car accidents, and drops in the value of their mutual funds. For them, the imperatives of the moment include health care, memberships to the health club, air bags, and financial advice. These imperatives become focal issues in our economy, and increasingly our economy is driving our culture. Our culture is coming to be defined more by economics than by values or politics.

Among economists, the word "values" does not even refer to principles or doctrines. It means prices—amounts of money that can purchase speedy travel, possessions, space to live in, and separation from bad smells and unpleasant tasks. The economists' definition of value is heard more and more, from the radio waves to private conversations. Economics holds a larger share of our public space than politics does, and more than culture does. It utilizes the advertising media, billboards, television, junk mail, new lingo in shops, and Internet banners. It shapes much of our culture, from decisions of what music to produce and market to decisions of which fashions to sell. It even shapes much of our politics, where economic growth is the supreme goal. Economics is ascendant. It fills the streets with cars and homes with furniture. Value means purchasing power.

With inexpensive prices as a guiding principle, speed becomes a new version of freedom, and cars and airplanes proliferate along with it; privacy becomes the new liberty, along with its larger homes; possession of plentiful devices designed to give us comfort and ease is the new American dream; and separation from toil, responsibility, and dirt becomes the new goal. The pursuit of such goals, dreams, and freedoms paves roads, strings telephone lines, creates housing developments, and alters the environment in myriad ways. They are the physical manifestations of the marketplace's new "values," and they represent direct change in nature as much as they represent each American.

They are also the activities that are shaping our personalities. We can see their results directly when we look at environmental change, but the values that lie underneath environmental degradation are at work on more than nature. They are at work on ourselves and on all aspects of the country. We can witness them in smog, climate change, poor drinking water, and the extinction of species. But we live them in urgency and fear and pressure. We can also witness these values in the fact that we have worked hard to restore many rivers, replanted some forests, built hospitals, and made many improvements in our surroundings. We live those benefits as well, with healthier bodies and educated children. All of these traits are the physical manifestations of our new national personality.

The New Manifest Destiny

In 1845, the editor of the *New York Morning News,* John L. O'Sullivan, wrote that it is the "manifest destiny of this nation to overspread the continent allotted by Providence for the free development of our yearly multiplying millions." When the United States did finally encompass its present contiguous area, some 30 years later, the growth ended 300 years of Spanish-Mexican control of many regions, eliminated control of some areas by France and Russia, and resulted in one of the largest countries in the world.[8] But these geographical achievements did not end the former colonists' drive.

Having reached California and purchased Alaska, the United States had taken the area between the Atlantic and Pacific Oceans by the second half of the nineteenth century. What goals would Americans turn their sights toward next? Physical expansion no longer had the same imperative it once had, since the country already stretched over many of the richest areas of the continent. Hawaii was still to be acquired, along with Puerto Rico, Guam, and other "possessions." But claims against Canada or Mexico were hardly a national obsession the way they had been once.

Americans could have been content. We could have continued to live on farms and ranches, to have large families to provide many hands to work the fields, and to live relatively local lives. But we are a country of explorers and conquerors. Our people had made themselves busy taking and then developing first the east coast, and then the west. Now if they could no longer conquer new geographies, they would conquer new economic opportunities and technologies. If they could no longer grow outward onto new land, they would grow instead in other ways, with new possessions and new opportunities.

Their next goal was to accomplish the same goals as those of the original settlers of the West but to do so within the boundaries already set geographically. The territoriality once satisfied by staking claim to a large isolated farm or ranch instead became the privacy of a large home in the suburbs, or of a room to oneself. The open plains were replaced with open roads that people could drive in the privacy of their own car. The physical wealth of the frontier was replaced with the physical wealth of consumer goods; the frontiers of electronics were pushed back the way the frontiers of the West once were; and the potentials of mass production were brought forth.

People were still seeking their own places, their own destinies, their own wealth. But they were seeking them from devices like light bulbs and phonographs rather than from open spaces. The light bulb gave them the evening and the night time, possessions that could not have been delivered by the ownership of more land; the phonograph gave them a musical existence that represented a place never before accessible. The automobile would multiply the physical freedom to reach new geographical areas that once came from the ability to stake out new land claims. Forms of economic security, like bank accounts and stock portfolios, would compete with older forms of security like the ownership of real estate. All of these devices opened new kinds of frontiers.

The "manifest destiny" to make the United States stretch the width of the continent became instead the manifest destiny to make the United States the fastest country in the world; to give its people the largest houses; to give them the fastest communication; to give them unique new possessions like the phonograph, the camera, and moving pictures. This new credo of manifest destiny was one that could be pursued in homes and offices instead of through colonization of vast distances. We are still pursuing that destiny.

In America today, it is an article of faith that each year more and newer technologies and devices will give us new opportunities for entertainment, comfort, travel, and new access to information. Our *Fortune* 500 corporations devote their activities to pushing back the boundaries of next year's equivalents of today's fax modems, steel-belted radials, and cell phones, which we are already setting aside money to buy even though we do not know exactly what they will be. Like the conquest of the West, it is an inevitability that our economy will generate new objects to buy and opportunities to explore. The desires they satisfy have forged our destiny, a destiny fulfilled not on an epic landscape, but in the mundane objects of everyday life.

Much of the rest of the world is now pursuing these destinies. People in China, India, or Brazil have not, through most of their history, assumed that their homes or possessions would grow larger, faster, more private, or more abundant. Some of them—the new "middle classes"—are beginning to expect such changes. As television programs from the United States have been transmitted to ever larger audiences around the world, American expectations have to some extent been transmitted as well. Now economic growth is starting to make this expectation affordable to larger numbers of people in other places, most notably China and India, the world's two most populous countries. America's hunger for such commodities as privacy and rapid transit is reaching and changing people in other countries, including ones with little tradition of privacy or fast-paced living.

Just as the first era of manifest destiny transformed places and trampled many people, including the American Indians who lived in the path of the juggernaut of European expansion, so our current trend has many harmful results. Its burden on nature may be as large as that of the original "taming" of the West—the conversion of grasslands and forests into farms, the overdrawing of freshwater, even the dust bowl, and the subsequent paving of cities and suburbs, pollution from factories, and legions of cars.

During the lifetimes of many people alive today, human existence has changed dramatically. Such new devices as refrigerators went from extravagances to commonplace; birth control became easily available; outhouses were abolished in most places. Telephones put us in touch with people far away. Travel became routine. People washed daily. Meanwhile, polio was almost eradicated; smallpox is almost gone; and radio became almost universal. Agricultural equipment reduced the number of people needed to raise crops. Factories absorbed the people who needed to find a different kind of work instead. Computers began to play games with kids the way playmates, neighbors, and parents once had.[9]

All of this is a physical reforming of the country. None of these things were among the goals of the founding documents of the United States. But they became our goals. They became more than our goals. They have doubtlessly affected our personalities, affected the length of our attention spans, the strength of our friendships, and many of the most fundamental, most personal aspects of our lives.

Where are these changes taking us? If our history has been driven in significant part by gradual, incremental change in our personal lives rather than by monumental public or political events, can we expect that our future will be increasingly private or individual? Will more of us live alone in the future? Are we going to travel over ever larger distances every day when we go to work and visit our friends? Are we going to move to new homes every third year instead of every fifth year? Some of the answers to these questions may be found by a look at our recent history.

Notes

1. Speed of information from Edward Tenner, *Why Things Bit Back* (New York: Vintage Books, 1996).

2. Robert Kaplan, "Travels into America's Future," *Atlantic Monthly* July, 1998.

3. Robert D. Putnam, "Bowling Alone: America's Declining Social Capital," *Journal of Democracy,* January, 1995; Robert D. Putnam, "The Prosperous Community: Social Capital and Public Life," *American Prospect,* Spring, 1993.

4. Bill McKibben, *The Age of Missing Information* (New York: Random House, 1992).

5. For up-to-date comparisons of book sales, see the *New York Times* bestseller lists and the number of books represented from each genre.

6. R.E.M., *Monster* (Burbank, Calif.: Warner Brothers, 1994); Lou Reed, *Berlin* (New York: BMG Entertainment, 1973 and 1998).

7. Alfred North Whitehead quoted in David Schiller, *The Little Zen Companion* (New York: Workman Publishing, 1994).

8. John L. O'Sullivan, ed., *New York Morning News,* 1845, quoted from a plaque in the Sonoma Barracks of Sonoma State Historic Park, California.

9. Bill McKibben, *The Age of Missing Information* (New York: Random House, 1992).

Alice never could quite make out, in thinking it over afterwards, how it was that they began: all she remembers is, that they were running hand in hand, and the Queen went so fast that it was all she could do to keep up with her: and still the Queen kept crying "Faster! Faster!" but Alice felt she could not go faster, though she had no breath left to say so.

The most curious part of the thing was, that the trees and the other things round them never changed their places at all: however fast they went, they never seemed to pass anything. "I wonder if all the things move along with us?" thought poor puzzled Alice. And the Queen seemed to guess her thoughts, for she cried "Faster! Don't try to talk!"

Not that Alice had any idea of doing that. She felt as if she would never be able to talk again, she was getting so much out of breath: and still the Queen cried "Faster! Faster!" and dragged her along. "Are we nearly there?" Alice managed to pant out at last.

"Nearly there!" the Queen repeated. "Why, we passed it ten minutes ago! Faster!" And they ran on for a time in silence, with the wind whistling in Alice's ears, and almost blowing her hair off her head, she fancied.

"Now! Now!" cried the Queen. "Faster! Faster!" And they went so fast that at last they seemed to skim through the air, hardly touching the ground with their feet, till suddenly, just as Alice was getting quite exhausted, they stopped, and she found herself sitting on the ground, breathless and giddy.

The Queen propped her up against a tree, and said kindly, "You may rest a little, now."

Alice looked round her in great surprise. "Why, I do believe we've been under this tree the whole time! Everything's just as it was!"

"Of course it is," said the Queen. "What would you have it?"

"Well, in our country," said Alice, still panting a little, "you'd generally get to somewhere else—if you ran very fast for a long time as we've been doing."

"A slow sort of country!" said the Queen. "Now, here, you see, it takes all the running you can do, to keep in the same place. If you want to get somewhere else, you must run at least twice as fast as that!"

—Lewis Carroll, *Alice in Wonderland,* 1897

CHAPTER ONE

Speed

The Fastest People Who Ever Lived

> I just bought a microwave fireplace. . . . You can spend an evening in front of it in only eight minutes. . . .
>
> —*Steven Wright*

Americans are the fastest people who have ever lived. We have fast food, one-stop shopping, one-hour photo developing, microwave burritos, fifteen-second television commercials, and a legion of time-saving devices. We wince when the car ahead of us in traffic slows down.

Measured mathematically, in miles of travel per day, we move faster now than ever before. Measured in calculations per second, we move more information than ever before. Our cities are built and rebuilt to serve our transportation needs. Our jobs are redefined to meet the needs and opportunities that come from more information. Our atmosphere is burdened with carbon dioxide. Our countryside is transformed into housing developments and industrial parks. All as a result of our higher speeds.

Our speed comes in many varieties. In offices, the goal of going faster is met by ever faster computers and, for some of us, frequent deadlines. On the roads, the desire for going faster is met by fast cars and straight highways. At factories, assembly lines are made more efficient and more prolific to produce more widgets in a year. What unifies these different kinds of speed is the common priority of more—making more, going more places, seeing more people, owning more, and doing more in a day or in a lifetime.

It is a country with high expectations for how much it can fit into a day or a year. Many Americans pack more activities into their daily schedules, more miles onto their odometers, and more possessions into their homes than their ancestors could have. Americans accomplish an amount of work and travel and experience never before accomplished in history. We are the envy of the world for the possessions and opportunities that we have, in part because we run a race and keep score in dollars per year and appointments per day.

High speeds not only add to the quantity of experience, they also change the quality of experience, and substitute different experiences. Fast transport takes people out of their own neighborhoods and into other neighborhoods and places, which is both a gain and a loss. Electronic access to information keeps us abreast of distant events while sometimes distracting us from local ones. Doing more enriches some people, but also makes many of us frantic. Going fast is one of the trade-offs of this era. It is the bargain that Americans have made with the clock, and we reap the benefits of that deal as much as we pay the costs.

These speeds have risen continually over the decades to reach new highs. People who in the early 1900s could travel no faster than a horse or boat could carry them would be shocked to see their great grandchildren's flurry of motion. The average American today covers at least 24 miles a day with motorized transport. Those with computers perform hundreds of thousands of calculations per second, putting slide rules and finger counting to shame. Much of modern experience can be measured in units of miles per hour, calculations per second, widgets per minute, points per game, and other figures of activity over time.[1]

Emotions like caring and hoping are sometimes carried out under pressure of time deadlines at work or travel schedules that require another appearance at the airport. The best basketball players are fast on the court and can virtually fly through the air. Schedules are packed. Some people are expected to relax faster. Messages are left on machines by people moving too fast to wait for an opportunity to actually speak to their colleagues and friends. FedEx and priority overnight mail services already seem slow since the advent of electronic mail.[2]

It is a regime of speed. To serve it, Americans have changed almost every aspect of their lives, from the people they see to the places where they live. We have even changed the physical state of our country. To make fast motion possible, we put asphalt where once there was tall grass, and motels alongside roads and parking lots that cover 153,730 square kilometers of America.[3] We have strung wires across the entire country, reaching every little corner, to let instantaneous telephone communication reach all of us. We have also left graveyards of automobiles, and massive environmental damage, from polluted streams to tens of billions of tons of carbon dioxide in the atmosphere.

Speed may be the greatest missed story of the century. History books have focused, correctly in part, on politicians, wars, and laws. Public policy documents have chronicled important changes in social spending, crime rates, interest rates, and many government affairs. But speed is even more fundamental. It changes everything we do. Its growing pace has taken us unaware, evading adequate coverage in our newspapers, which tend to capture the events of the moment instead, and evading conscious attention in many of our own lives because we are too caught up in the race to see that the pace has risen—gradually but continuously.

We are at the fastest moment yet in history. The greatest test of whether we can build the kind of world that we want may be the test of whether we can cope with this speed, or even whether we can use it to serve our values and goals. The greatest environmental test that we face may be to change our relationship with time, and the pollution and disruption that surround rapidity. In each case, the first step is to see our changing speeds and the effects that they have on all of us.

Let the Race Begin

The year is 1914. The place is a street corner in Cleveland, Ohio. This corner is suddenly home to the world's first traffic light. Finally we have enough cars to require us to take turns waiting. The light has no yellow, but shines a bright red and green. At the beginning, puzzled drivers pass by occasionally, probably wonder what to do, and then drive on. Yellow will be added six or seven years later, on a traffic signal in Detroit, Michigan. These greens, reds, and yellows are the beginning of a new era of speed.[4]

The era wakes up slowly. Americans get around relatively little at the beginning of this time. Few people have cars; the cars they do have move slowly; and even though airplanes have existed since 1903, they are not available to the masses.[5] Using all forms of motorized transport—cars, buses, motorcycles, trains, and airplanes—Americans cover about 3 miles in a day on average (see figure). This seems about right for a time when the Roaring Twenties had not yet begun. In one year, that amounted to a distance long enough to get a typical person from New York to a point about halfway across the country, toward San Francisco, if he had gone in a straight line.[6]

Already by the 1920s, though, American drivers had learned speed-reading to take in the messages on billboards that they had started to fly past at record speeds. On road signs, abbreviations increased their importance in our language: motels received names like "Ko-Zee," "Sleep-U," or "Tour-Rest." Restaurants featured "Bar-B-Q." The advice to stop for food was reduced to "Eat." This historical change did not only take place on the roads, but also in homes and offices. A Dupont advertisement in 1922, for

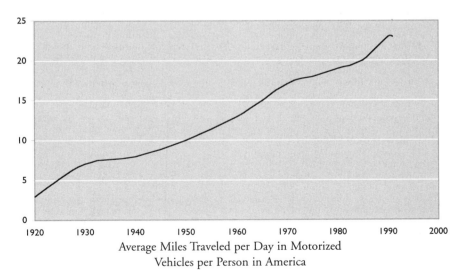

Average Miles Traveled per Day in Motorized
Vehicles per Person in America

Sources: Calculations by Cliff Cobb based on data from the U.S. Bureau of
the Census, *Statistical Abstract of the United States,* historical editions;
American Public Transit Association, *Transit Fact Book;* Eno Foundation
for Transportation, Westport, CT: *Transportation in America,* various years.

example, celebrated the time-saving chemical engineer: "It is he who has helped make your minutes as long as your great grandfather's hours."[7]

The 3-miles-a-day speed did not last. By the end of the Roaring Twenties, jazz music had freed many Americans from the fixed patterns and paces of older music, and mechanical transport had freed many of them from the constraints of space. By the time of the stock market crash of 1929, the average American used mechanized travel to cover fully 7 miles a day—every day.[8]

In just one decade, Americans' average travel had more than doubled. For the first time in history, the residents of the North American continent moved a physical distance long enough to reach from coast to coast. At the 1930 average annual travel distance of 2,413 miles, a typical motorist who set out from New York in a straight line could have seen California during the same year. For those who chose not to drive in a straight line, they could visit cities and towns and parks in their own state that previously had been beyond their reach.

The Great Depression slowed our increases in motion, however. The 1930s were the only decade since the time when cars had begun to be mass produced that had seen little increase in national speed. By 1940 our average motorized distance was only one mile more than it had been in 1930. Even so, people were going fast enough to change traditional ways of life. The average distance covered in a day with motor vehicles around 1940 was over 8 miles—a length that would take a typical person more than two and a half hours to walk. People could use their cars to leave their communities or to seek employment or entertainment farther from home than they had at any time in history.[9]

By midcentury, this freedom had increased. Mechanical travel carried Americans 10 miles a day. The United States had come out of World War II on top of the world, and it would invest its strength in speed. According to journalist Phil Patton, in the mid-1950s boosters of interstate highways argued that a highway "turned space into time." With a fast road you no longer talked about how far you had to go, but about how long it would take you. "How far away is it?—Three hours." If roads did not create new places to go, what they did was equivalent in effect. They made it possible to go to more places in the same amount of time.[10]

Yet even at this time of U.S. dominance, the country only moved at about 40 percent of its speed today. Its citizens did not even go one-half mile per hour, on average. But their eyes were fixed firmly on the highways that were to carry us so far and so fast. President Eisenhower spoke of how highways would provide "greater convenience, greater happiness, and greater standards of living." They were part of the vision of the American way of life. New highways, it was argued, "would shorten the time it took to get to the suburbs, effectively expanding the space of the city, trading off space and time in the traditional American manner." By the end of the 1950s, the suburban population of the country equaled that of the cities themselves.[11]

By 1960, average U.S. motorized distance traveled in a day had reached 13 miles. During that decade, in many states, highway building accounted for well over half of all public capital spending. We were buying speed. By 1970, the national motorized speed was 17 miles a day. In 1985, it reached 20 miles a day. And then it was 23 miles a day in 1990.[12]

Without new census data, it is not clear exactly how fast our motorized motion is today in the United States. In a few crowded places, like parts of Los Angeles, this motion may even have declined because of traffic congestion. But for the country as a whole, data on car sales, gasoline sales, and living patterns suggest that our motion has risen. It has almost certainly reached 24 miles a day—eight times the amount in 1920. At an average of more than 24 miles a day, we have passed the one-mile-per-hour mark, measured over every hour of every day, even including the time we spend asleep.

The Speed of Information

In the 1950s, a Tennessee congressman and senator named Albert Gore shepherded through Congress legislation to create the interstate highway system. It was legislation that would form the country, both figuratively and literally, creating a national mentality of mobility and also a world of asphalt. In the 1990s, another Tennessee legislator, Albert Gore Jr., pushed for equally formative legislation in Washington: a commitment to the "information superhighway."[13]

In many ways, the excitement of the "interstate" has been given to the "Internet." Albert Gore Sr.'s roads cut new ground that was as unknown

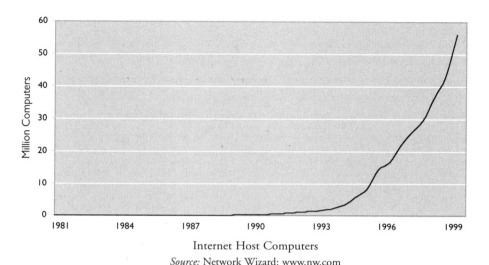

Internet Host Computers
Source: Network Wizard: www.nw.com

then as cyberspace is today. Both have changed the ways that Americans spend their time as well as their relationships with families and friends, the job opportunities they have, and their experience of the world. Highways allowed people to visit family members and friends who lived increasingly farther away. Just in the past few years, the Internet has made it possible to trade electronic messages and information at low cost with people who live far away, as well, in this case even on other continents. Both technologies have allowed us to be in touch more often with our friends, colleagues, and relatives.[14]

Increasing speeds of information have allowed more people to work at home by "telecommuting." This has changed professional relationships and let some parents spend more time with their children. It has changed our workplaces and our homes. Increasing speeds of information have become an educational tool, with CD-ROMs used in schools and homes, encyclopedias that fit on one disk instead of a long, thick row of books, and "Web surfing" as a new hobby and research tool. Much as people became able, a few decades ago, to roam and explore neighboring counties and states, they have now become able to roam through information sites located around the world.

The movement toward speed of information has roots much farther

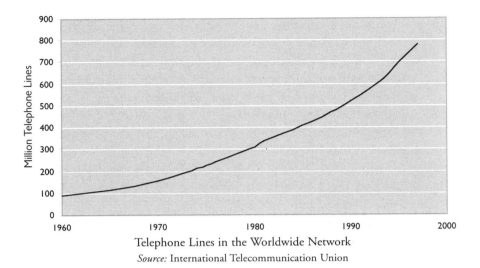

Telephone Lines in the Worldwide Network
Source: International Telecommunication Union

back in history, though. The invention of the telephone in 1875 permitted, for the first time in history, instantaneous responses to someone more than a few yards away. This was considered a crucial development at the time, and was the sole activity of Columbia, the symbol of America, in John Gast's centennial portrait *American Progress*. Since then, telecommunications have made dramatic strides. In 1951, long-distance telephone service was introduced to American homes. This meant that people could chat with family members and friends from across the country or the world. It reduced people's isolation by letting them hear distant voices. We now take these calls for granted. It would never occur to us that we might not be able to talk to our friends without undertaking a journey of days or months.[15]

The year 1951 also marked the unveiling of the first computer in the United States—the UNIVAC (Universal Automatic Computer), made by IBM. It was installed at the Census Bureau. In 1953, a high-speed printer was connected. Even so, it did not have the speed or software to carry out many of the functions that we use every day today, even on our home computers. By 1975, we had desktop computers, although not in our homes. There were probably fewer than 2 million computers in the world in 1980, most of them mainframes not accessible to ordinary people.

Today, many people in America take computers for granted, and could not work, write, communicate, or even feel at ease if a computer were not nearby. Not many people today would even consider undertaking a long writing project with pen and paper.[16]

These computers have even speeded up our shopping. After all, who wants to wait in line? The Universal Product Code appeared in 1973 and allowed electronic scanners to read the prices of items in the supermarket and record them automatically. Looking around in supermarket lines today, it is possible to see frustrated people wishing that the check-out clerk would move faster. But how much worse would their mood be without the scanner to move things along?[17]

Communication through the mail has sped up as well. In 1971, Federal Express opened for business, carrying letters and packages across the country in one day, or overseas in two. Since then, it has been rivaled by competition from several challenging companies and the U.S. Post Office's overnight mail. But even overnight delivery was overtaken within just a few years. In the mid-1980s, fax machines became commonplace. Overnight mail did not become obsolete, but it had been bested after only about fifteen years.[18]

The race for fast, flexible information did not stop there. In the late 1980s, cellular telephones entered the market. People could talk to each other from their cars, from the grocery store line while they waited for the scanner, or from the restaurant. Entire industries have sprung up to produce portable telephones and meet the needs of mobile communications. By the mid-1990s, the information industry had made electronic mail widely available. Suddenly even faxes seemed awkward.

The software industry has since added "simultaneous chatting," in which computer users are notified when their friends or colleagues turn on their computers across the country and are able to send messages to each other that are received instantaneously. It is the computerized version of a telephone conversation.[19] Meanwhile, Internet shopping has spurred rapid growth in overnight delivery services that fill orders, and so has fanned the fires beneath FedEx and UPS once again.

The information industries will not stop there. Soon, many of these technologies are likely to merge, with cable television lines joining with telephone lines, Internet access, and other functions, and with satellite dishes providing similar services. Integrated computer, television, film, and telephone services will allow people to mix all of these functions together. And the industry will continue to evolve from there.

The Culture of Speed

As transportation, information, and other aspects of American life have quickened, their speeds have become part of the culture. Attitudes and services and institutions have changed to suit America's rising speeds, and, in turn, those cultural changes have reinforced the tendency to go fast.[20]

America has built an infrastructure of speed—not only of roads and planes, but also of plentiful coffee, and even of social encouragement for speed. It includes widespread social acceptance of hurriedness. Colleagues boast over who got the least sleep the night before. To have less sleep is taken as a measure of productivity or fullness of life. Tired people seem busy and, therefore, productive and hard working. Meetings are held at airport conference rooms by people too busy even to visit the city that they have just flown in to before they fly back out again. Those people who hurry appear to be in demand. They may even seem popular because they cannot be in enough places or talking to enough people at once.

These cultural institutions not only make speed more possible and more comfortable, they also make it more respectable, even enviable. They build a cycle, where those who go fast receive the trappings of speed and the encouragement to go faster. In the office, busy people may receive an assistant or a secretary to help them get more done. At home, they may hire a maid to clean or a carpenter to do some work on the house that they would have done themselves if they had more time. Their status (and the cost of their time) justifies these expenditures.

Money plays a role in this cycle. Those who can get the most done in a short time have an opportunity to earn more. Salespeople and others who are paid on commission see this directly. It applies to many other workers indirectly because they receive promotions and pay raises faster if they are more productive. Many of them then invest part of their earnings in the equipment, people, and education that they need to work even more quickly and effectively in the future. They contribute to the infrastructure of speed, both physically through their spending and culturally through the example that they give to their friends, their children, and their colleagues.

Even for whole countries, this is true. The United States' total earnings, its gross domestic product, is the highest in the world. It is no coincidence that it is the country that invented the airplane, the automobile, the telephone, the computer, and the assembly line. It was Henry Ford who invented that first assembly line, and he did it while he was trying to build cars faster. (His invention cut the time it takes to make a car from fourteen person-hours to just two.) Ford chose, famously, to pay his workers enough to allow them to buy their own Ford cars, giving them a stake in speed and completing the cycle. Meanwhile, much of the rest of the world tries to imitate the wealth and some aspects of the culture of the United States.[21]

The money and status of having accomplished much in a short time are addictive. People who have done so can afford expensive homes and foods, and they have to remain productive to finance those belongings and maintain their ways of life, or to improve upon them. This often requires long hours of working hard, and sometimes working frantically, because people want to return home to their beautiful, expensive, homes.

This cycle is not just for the rich. The simple needs of paying rent or the mortgage for any home, affording to have children, and bringing home food now require many families to have two incomes, and some people to have two jobs. For two-parent families, with both parents working, it is a race to take care of the family as well as two jobs, the home, and all errands and responsibilities. For single parents, or other people with large responsibilities, keeping up may be even harder. It is no wonder that people feel the need to move fast, if they are to keep up with it all.

Social critic Susan George argues that speed is about prestige. Farmers, she says, are tied to the speed of the seasons and the day—and they are the least prestigious. The slowest are least prestigious everywhere, she says. They are rooted in a particular place and work with the seasons, which are slow by definition. The fastest, then, are the most prestigious. Stockbrokers who shout out urgent bids on the floor of the New York Stock Exchange make a good income, wear expensive suits, and have prestige.

People who face many deadlines, sudden emergencies or crises, and considerable pressure in their work often have prestige: doctors, lawyers, senators, chief executive officers. The fast, prestigious people are the ones who donate to political campaigns, belong to influential networks of people, and live in large houses. The speed/prestige pattern gives politicians and chief executive officers, as well as doctors, another reason to be in a hurry.[22]

If slowness gives low prestige, then people who move slowly, like the elderly, or those too poor to afford an automobile, will have little prestige. In slower-moving countries and traditional societies, the elderly are held in reverence. The oldest are the wisest, and often play the role of a chief or a leader. But in America, moving slowly means having less ability. Here, power is for the taking by those mobile enough to reach the voters or climb the corporate ladder. This changes our society dramatically, giving decisions to different people who will make different choices and serve different priorities. They will not tend to maintain slow-paced, rural places where people sit in rocking chairs on the front porch. They will tend to transform those places into towns and subdivisions that people can speed through in their cars on their way home from work.

People kept slow by poverty will be stuck in a trap of poverty if speed is what is required to break out. Not having a car, or a computer to access the World Wide Web, will be one side of the cycle that prevented them in the first place from having enough income to buy a car or computer. The saying goes that the rich get richer. Now maybe that should read that the fast get faster. It is the speed-based version of environmental and economic injustice.

The country with the most cars, computers, and satellite dishes also has the most power. Just about 8 percent of the world's people have a car at their disposal, and only around 3 percent have access to a personal computer. The United States dominates those populations. Not coincidentally, it is also the country with the most military, economic, and, some would argue, cultural force. It puts its television reruns on screens in homes around the world, and its music on the radio waves of distant places. The United States can do so because it has a large infrastructure of technologies and specialists to move information, pictures, and people quickly. It

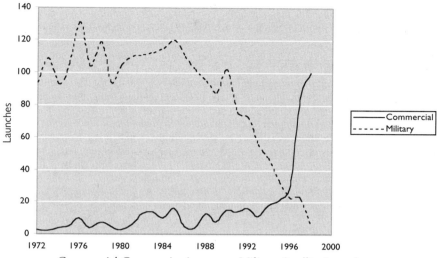

Commercial Communication versus Military Satellite Launches
Source: Compiled by Worldwatch Institute: 1972–97 from Teal Group; 1998 from
Heyman. See Worldwatch publication *Vital Signs 1999* for more information.

has trained its people to use computers and has made flying in airplanes commonplace. Its airplanes and computers are among the machines that the United States uses to spread its culture and its economy.

According to Susan George, "the fastest have, throughout history, been the most powerful." They can capture the dynamism of their motion or knowledge or capabilities to make the world the way they want. People can't really organize to slow things down, she says. But people can organize to speed things up.[23] Because of the relationship between speed and power, we have long sought to institutionalize speed, both in governmental affairs, in which the elimination of bureaucratic inefficiency has been in vogue over the last two decades, as well as in the private sector.

Speed and America's most cherished institutions have been tied together. The motor-voter law, for example, allows people to register to vote when they get their driver's license. When people do vote, many of them base their decisions on information from the television news, which notoriously spends very little time on any one topic. If voters took more time to read at length about candidates and issues, their votes might change. The messages of their politicians might change as well, since those messages would no longer have to fit in eight-second sound bites. But that would be contrary to our culture of speed.

Many American institutions serve people under pressure. Even our restaurants and shops are geared to speed. Some restaurants are designed specifically for it. The drive-up window, for example, and the "prepared in five-minutes-or-less 'business lunch'" at some restaurants, justify fast-paced living by institutionalizing food for people in a hurry. For those who call for delivery, orders that take more than thirty minutes will receive $3 off. Many people plan their schedules knowing that if meetings run over they can grab some quick food. If this speed were not available, people might see their days differently and revise their plans so that they would not have to go hungry.

These changes reach back into American history. America's first shopping mall opened in 1950, and shopping malls have proliferated across the country since then in part because by putting many merchants under one roof they saved busy people the time of searching out the right shop. The first mall was Northgate Center, located north of Seattle and accessible only by automobile. It included thousands of parking spaces, about eighty stores, a movie theater, and a service station. Northgate Center was put to shame, however, by the mall that opened in 1992 near Minneapolis, Minnesota. The builders of the Mall of America projected that it would attract more visitors each year than Mecca or the Vatican.[24]

Other institutions were created even farther back in history to serve people on the move. The world's first "motor hotel," or "motel" opened on December 12, 1925, in San Luis Obispo, California. It provided a dignified way to sleep on the road. It resolved our needs for sleep with our culture of covering many miles per day. This 160-person accommodation led the way for what would become, by the mid-1960s, a massive fleet of about 40,000 such dwellings dotting the highways of the entire country. Motorists could arrive in the evening and be gone by early morning.[25]

The motel gave birth to a whole culture of fast travel. Motels have added many miniature icons to America's culture. These include small, disposable bars of soap; miniature bottles of shampoo; neon lights; the phrases "no vacancy," "open 24 hours," and "self serve"; sheets slept on and then washed after only one night; and the full set of motel and coffee shop services. These amenities changed the experience of travel, changed the nation's environment, and took their place in a highway subculture that is distinctly American.

The priority of speed reaches into many other minute accessories that make speed desirable as well. It includes such little luxuries as inflatable airplane pillows, suitcases with wheels, and espresso booths on the street to wake up the sleepy without requiring them to wait in a coffee shop or grind beans. For those men who do not give themselves enough time in the morning, the electric razor can help. It can even be used in the lavatory of the red-eye cross-country flight, so that a busy person can arrive early in the morning, fully groomed, thousands of miles from the previous night's dinner. For women and men there are portable stereo systems, laptop computers, gadgets to press wrinkles out of clothing, and whole catalogs full of similar accessories.

The culture of speed includes entertainment. On June 6, 1933, the first drive-in movie theater opened, in Riverton, New Jersey. It came at a time in history when many people liked to spend time in their automobiles, even when they were stationary. By 1960, there were 4,700 drive-ins in the United States. They combined Thomas Edison's moving pictures with the ability to move oneself easily to the outskirts of town where there was room for an outdoor theater. More recently, since the passing of most drive-in theaters, entertainment has offered speed in other ways. Music television does not demand more than about three minutes of attention from any viewer. The average length of a television commercial has been shortened from thirty seconds to fifteen to keep up with our rising speeds and with viewers who will not pay attention for more than a quarter-minute.[26]

Some of our institutions of speed even serve as national gathering places. We do not have very many common spaces, where people can gather and watch each other and gauge their own place in society. The ones we do have include "Main Street," where drivers once "cruised the strip" in the 1950s, and the Internet, which is the newest common space. These institutions that were designed to let us move quickly now also serve to let us feel connected to other people, look at them, and share gossip and news. Without them, Americans would be more isolated. The highways and the telephone lines are the arteries that connect people to each other. We are dependent on speed, and on the devices that give speed to us, even for human contact. Sometimes we use speed to fight isolation, and this is a powerful benefit that rapidity has given to us, although it may also be a technological crutch that we use to make up for a way of life that some people find does not fulfill their needs for companionship.

The culture of speed gives us some other benefits as well. Among them is a belief in an entitlement to emergency health care. We have all grown up believing that if we become ill or need emergency attention then a call to 911 should bring paramedics to our door fast enough to rescue us from even a heart attack. The roots of this ethic reach back to 1928, when Julian Stanley Wise organized the first American volunteer rescue squad trained for emergency medical situations. These emergency medical technicians have since become a force of more than half a million throughout the world. Their ability to respond quickly to emergencies with ambulances and other motor vehicles has saved lives, promoted health, and made people feel more secure for years. It has taken speed right into our health care system and into our sense of safety.

The Effects of Speed on the Environment

While speed has many benefits, many of its burdens fall to the natural environment. It is nature that provides the raw materials for fast vehicles and fast assembly lines. It is nature that absorbs the exhausts and leftover materials from the use and construction of machines for speed. Nature provides these materials and services free to consumers and companies who take advantage of its ability to absorb wastes, provide fuels, accept asphalt, and wash away grime and pollution. But the environment suffers the consequences in the form of smog and polluted streams and open pit mines and a changing climate, as well as in other ways. In a sense, nature pays the bill for the pleasure and the power of people who travel at high speeds and build computers that do high-speed calculations.

In some cases, people's ability to move and work quickly can be turned to the advantage of the environment. By using increasingly powerful computers, organizations can keep track of what is happening to the environment, and can learn how to manage it better. By using airplanes and cars and boats, environmental managers can survey forests and coral reefs and grasslands that would not otherwise be accessible. Tourists who do not normally leave their communities can go out and experience national parks and increase their appreciation of nature. On the whole, however, these activities to protect the environment are overwhelmed by the scale of activity that causes harm and arises from Americans' need to drive every day, have products delivered quickly to their doors, and use the environment in myriad ways.

Computers are an example of the trade-offs of high-speed technology and the environment. Computers allowed researchers to predict climate change and begin to promote policies to avoid it. Computers share information around the world that is crucial to environmental protection. They help many companies design manufacturing processes that reduce waste and conserve resources. But Silicon Valley, where many computers are designed and built, has the highest concentration of hazardous-waste cleanup sites in the United States. Solvents commonly used in chip making, such as ethylene glycol ethers, are suspected of causing reproductive health problems. Trichloroethylene (TCE) and 1,1,1-trichloroethane, which are used to make and clean electronic components, are both linked to serious health problems.[27] Worldwide, computers were already consuming about 240 billion kilowatt-hours of electricity a year by 1993, as much as the entire annual electricity use of Brazil.[28]

Cars offer many trade-offs as well. People find them so desirable and valuable that they are a central part of most Americans' daily lives. Meanwhile, the area of the United States paved over by roads and parking lots is 153,730 square kilometers, a huge loss of nature. This is almost as much as the combined area of all U.S. National Parks: 191,501 square kilometers. Cars are one of the leading producers of greenhouse gases, smog, particulates, and volatile organic compounds. Their manufacture uses many harmful chemicals and produces many wastes.[29]

The more we increase our speed, the more we encounter these trade-offs as our pollution and disruption of the environment increase. Wolfgang Sachs of the Wuppertal Institute in Germany points out that "the average car, which consumes 5 liters of gasoline at a speed of 80 kilometers per hour will not just need 10 liters when it runs up to 160 kilometers per hour, but 20 liters. The high-speed French TGV train and the German

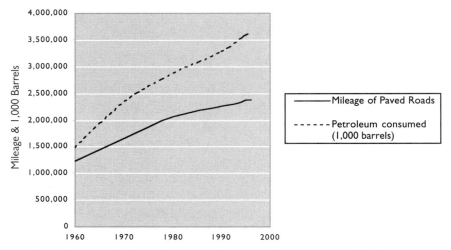

Paved Roads in the United States and Petroleum Consumed on U.S. Highways
Source: U.S. Department of Transportation

ICE consume not just 50 percent more when they jump from 200 to 300 kilometers per hour, but 100 percent more."[30]

The results of our motion are tangible. A gallon of gasoline weighs almost 7 pounds, of which about 6 pounds are carbon. When burned in the car, the carbon combines with oxygen to form carbon dioxide. The weight of the oxygen plus the weight of the carbon is about 22 pounds of carbon dioxide. This weight is far more than the weight of each gallon of gasoline that spawned it, but this is the weight that is lifted into the atmosphere as we drive. It is the atmospheric consequence of our fast travel.[31]

These effects on the environment of moving fast and getting information fast can be ameliorated by increasing the efficiency of technologies for speed—but only if we make investments in efficiency, and even then, only in some cases. Sulfur oxides and nitrogen oxides are pollutants that can be removed by catalytic converters in cars and scrubbers in smoke stacks. But carbon dioxide cannot be removed by either of those devices. Only investments in the development of really new technologies, like hydrogen-powered cars, might offer a chance of reducing our climate-changing carbon-dioxide emissions. Even hydrogen and electric cars emit carbon dioxide in a sense, however, because fossil fuels are burned to produce the hydrogen, electricity, or other inputs that they use. Carbon dioxide is not merely a by-product of transportation the way that sulfur and nitrogen oxides are, but rather is integral to our mobility.[32]

Driving is an act of putting carbon dioxide into the atmosphere to get us around town. If we stood with a shovel in the back of a car scooping carbon dioxide into the air to make the wheels turn, the way people once shoveled coal into the furnaces of moving trains, then we would understand directly. We would see the other side of town in terms of pounds of greenhouse gases shoveled into the air. Just like we now say that Cleveland is "three hours away," using time instead of distance, we could also say that Atlanta is "250 pounds of carbon away," using weight instead of distance as well. With those descriptions, climate change would appear as a social trend—a human behavior. Otherwise, it seems an act of nature, or one that happens independently of us as individuals.

Moving information quickly has many values, and many people have to do it today in order to support themselves and their families. Getting

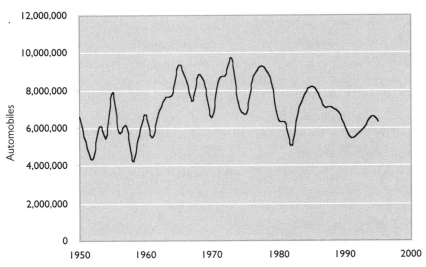

Automobile Production in the United States, 1950–95

Source: American Automobile Manufacturers Association, *World Motor Vehicle Data* 1991, 1993, and 1996 editions.

around town is an unavoidable need in our society and economy. But car-rying out our daily activities means taking part in a way of life that will soon force American society to come to terms with the environmental effects of this historical change of speed.

Wolfgang Sachs says "generally speaking, the ecological crisis can be read as a clash of different time scales; the time scale of modernity collides with the time scales which govern life and the earth." He goes on:

> Every year, the industrial system burns as much fossil fuel as the earth has stored up in a period of nearly a mil-lion years. Within a second, in terms of geological time, the planet's reserves are about to vanish in the fireworks of the industrial age. . . . It is obvious that the rate of exploitation of non-renewable resources is infinitely faster than the processes of sedimentation and melting in the earth's crust. Industrial time is squarely at odds with geological time. It is probably not an exaggeration to say that the time gained through fuel-driven acceler-ation is in reality time transferred from the time stock accumulated in fossil reserves to the engines of our vehicles.[33]

This is not only true of fossil fuels, but of many environmental prob-lems. This pattern applies to the excessively fast logging of forests that have grown over centuries; the sudden destruction of slow-growing coral reefs by fishing practices designed to bring in fish more quickly; the draw-ing down of underground aquifers that filled slowly over time; the sudden trampling of soils that have accumulated over the years by cattle bred and fed to grow quickly; the pollution of streams, soils, and air; and other problems.

According to Sachs, "processes like growth and decay, formation and erosion, assimilation and regeneration, selection and adaptation follow rhythms of their own. Pushed along under the fast beat of industrial time, they are driven into turbulence or destabilized. The speed of capital accu-mulating is at variance with the speed of nature regenerating."[34]

Rainfall returns about 2.4 billion gallons of water to the Ogallala aquifer in the Great Plains every year, but human beings withdraw about 20 billion gallons every year for irrigation and household and industrial uses. Nature creates about 0.4 billion tons of topsoil every year worldwide, but under our agriculture and land management, somewhere in the area of 25 billion tons are lost to erosion every year. Prior to the industrial revolution, it took the world about 30 years, on average, to lose 1 percent of its forests, but now the world loses 1 percent of its forests every year. During the past 65 million years, between 1 and 10 species became extinct, on a rough average. Conservative estimates now put the number of extinctions at 1,000 to 10,000 ever year. These are among the many collisions of speeds happening in the environment today.[35]

This collision of speeds takes place in the economy of food as well. In Sachs's words, "also in animal raising and plant cultivation, the natural rhythms of growth and maturation are considered too slow by the industrial mind. Enormous resources and ingenuity are brought to bear against the time inherent to organic beings to squeeze out more output in shorter periods. Cows and chickens or rice and wheat are selected, bred, chemically treated and genetically modified to accelerate their yield."[36]

Some of these activities are indispensable. Without the breeding of faster-growing varieties of grains and the application of more fertilizers, massive famine would have overtaken the Indian subcontinent in the late 1960s and other places many times since. Without rapid transportation, emergency medical teams could not have saved nearly as many lives as they have, people could not have gathered together to solve social problems, cultural activities would have been limited. Even the people who try to solve environmental problems need fast transport, readily available food, and very fast communications to do so. This is part of the irony of our time.

Airplanes, for example, have remarkable benefits, and are used heavily by people who work for environmental organizations. But those benefits have come at a high price. Airplanes are the most energy-intensive means of carrying people and cargo. On U.S. planes, carrying a passenger 1 kilometer takes 3,100 Btu; even by automobile it takes significantly less— 2,200 Btu. By intercity rail it takes only 1,500 Btu; and by intercity bus, only 600. But no other technology would allow us to link our culture together with those from countries across the world, or to collaborate with distant colleagues in person.[37]

The trade-offs between use and nonuse of a technology like air travel are complicated by changes in the efficiency and cleanliness of such technologies. Over time, for example, the energy efficiency of air transport has improved. Between 1973, the year of the first oil shock, and 1984, the amount of fuel American jets used to carry each passenger 1 kilometer decreased by 70 percent, partly from the use of wider-body jets with more passengers on each flight and from more-efficient turbines. Boeing's newest airplane, the 777, has a wider body and wingspan than any other plane and two engines instead of three or four.[38]

Despite the improvements in efficiency, though, jet fuel use has climbed steadily because the total distance flown has increased. In the whole world, only 472,000 tons of jet fuel were produced in 1950. But by 1970, that figure reached 81 million tons and by 1990, 156 million tons. That amounted to more than 10 percent of the world's consumption of transportation fuel.[39]

Americans lead the world in airplane fuel consumption, which increased from 187 kilograms of jet fuel per person in 1970 to 258 kilograms in 1990. Canada follows, with its consumption growing from 75 kilograms per person in 1970 to 123 in 1990. No other country of major

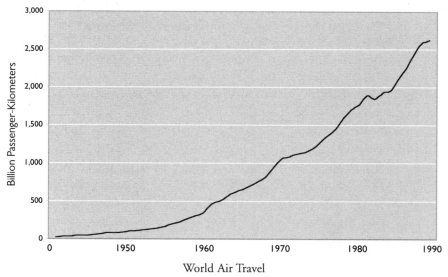

World Air Travel

Source: International Civil Aviation Organization. Figures for 1950–69 exclude former USSR, and 1998 figures are estimates by Worldwatch Institute based on ICAO projections.

size comes even close. Australia was using 75 kilograms per person in the mid-1990s; the United Kingdom, 55; France, 24; Japan, 23; Germany, 19; and Bangladesh, just 1.[40]

Along with that consumption has come pollution. In 1990, the burning of jet fuel produced about 550 million tons of carbon dioxide, 220 million tons of water (which acts like a greenhouse gas when it is high in the atmosphere), 3.5 million tons of nitrogen oxides, and 180,000 tons of sulfur dioxide, according to the International Energy Agency.[41]

What matters most about this pollution is the altitude at which it is released. At least 60 percent enters the air more than 9 kilometers above sea level, and as a result will stay in the atmosphere some 100 times longer than if it were released at ground level because it has fewer other molecules to react with. Furthermore, the concentrations of those gases—especially nitrogen oxides—is ordinarily low at such heights, so the additions cause dramatic change in atmospheric composition and contribute to the greenhouse effect. And they can lead to increases of tropospheric ozone, which also adds to greenhouse warming.[42]

At lower heights, nitrogen oxides undergo different reactions, and they play a part in chemical processes that break down ozone molecules, possibly contributing to stratospheric ozone depletion. Even the water released by air traffic, which would be harmless near the ground, causes disruption at high altitudes. It freezes in the cold upper sky, and the ice crystals allow sunlight through but then trap the energy radiating outward from the earth, thereby adding to global warming.[43] Yet, in 1998 the nation's airlines carried about 600 million passengers, up almost 25 percent since 1993.[44]

Similar losses and gains exist in our homes. Clothes washers, dryers, and dishwashers all let us clean up quickly. But what is ironic is that their speed dirties the atmosphere and the rivers. They transform one kind of dirt into another. Mud comes out of our clothes faster than it would if we had to wash them by hand, but sulfur dioxide goes into our air. Our dishes become squeaky clean, but our streams run with foam from detergents. Our clothes dry much more quickly than they would on a clothesline, but our rainfall is disrupted by climate change. On average, the use of electricity translates into the emission of 1.5 pounds of carbon dioxide per kilowatt-hour, and of sulfur and nitrogen oxides and other chemicals.[45] Those emissions are a reflection of our rush to clean.

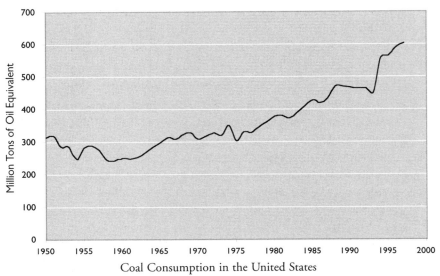

Coal Consumption in the United States

Sources: United Nations, British Petroleum, U.S. Department of Energy,
Journal of Commerce.

Given this situation, maybe our notions of ourselves as clean and well groomed are incorrect. With a full accounting of our cleanliness, we are slobs. We put a volume of sulfur, nitrogen, particulates, and volatile organic compounds into the air that people from earlier times in history could never have, using their hands and their backs. We use fast machines to make our homes and offices cleaner than almost any habitations have been in history, but we degrade nature faster than anyone has before in history. It is a trade-off. Cleanliness inside in half the time for disruption outside on a time scale never before seen in history.

Some people have hoped that the trade-off of our era would be between old, polluting technologies and newer electronic devices and electronic communications that would put an end to pollution. They hoped that the data highway could be traveled with no noise or exhaust fumes. But it turned out that the data highway requires major equipment and uses particularly high-grade minerals. According to figures from the Wuppertal Institute, no less than 15 to 19 tons of materials are consumed by the fabrication and life cycle of one computer, while for an average car it takes 25 tons. The computers that seemed so much less resource-dependent may place a burden on the environment similar to that of a car, and have shorter lifetimes of use.[46]

In the 1980s, many people voiced the hope that consumption of paper would decline when information was stored on computer disks and computer screens. But paper consumption rose as computer printers made it possible for people to generate paper copies more easily than they ever could have with typewriters. One conservative estimate put the world's consumption of paper, in 1992, at 230 million reams, or 115 billion individual sheets.[47]

Many people also hoped that electronic communication would substitute for the traffic of automobiles and airplanes, making those old, large devices less necessary. In some cases it has; but it has also stimulated new traffic by extending the network of people who we work with and visit. The number of miles driven in cars and flown in planes has risen steeply, and the number of cars and planes produced has risen as well, far more steeply than population growth. And it has done so during the time when computers spread across the United States, and when the Internet linked us together.

Industrial Speed

For most of history, economic and cultural change could spread no faster than a horse could run or a ship could sail. Later, it could spread by the speed of a car or an airplane. But now it spreads at the speed of the electric pulses in our televisions, computers, and telephones. Nothing, however, matches the speed or magnitude of change that flows with the electronic transfer of capital.

It is now possible for $1 trillion to change hands in one day, with electronic trading schemes and urgent transactions by brokers of currencies and equities. At a corporation's initial public stock offering, massive quantities of capital can flow into it in a matter of hours. If a country appears to become an investment risk, capital can flow out of it in hours, changing the entire economy of that country. This story has played itself out recently from Indonesia to Siberia. With the money, no physical change takes place—only the movement of information contained in electronic financial accounts. But those accounts, in turn, cause major physical, social, and economic change.

The connections between speed and capital are deep. It was the construction of railroads in the 1800s that first required the development of modern financial markets and of regulatory agencies and central banks, because large capital outlays were needed to build these huge machines that stretched for hundreds of miles. As a result, it took the precursors of today's huge and fast flows of money to make possible the building, not of financial speed, but of past physical speed—rail, roads, cars, and airplanes.[48]

Today, electronic transfers of money forge the invisible pathways that will be followed by construction workers who build new offices, apartment buildings, airports, and all aspects of our surroundings. First goes the financing, and then follow the natural resources, the labor, and eventually the customers. The faster those transfers of money can occur, the faster the labor and natural resources can be mustered to build new monuments to our prolific economy. And investment finance has been getting faster. As our pension funds grow, as our GDP grows, more of us become able to invest in the companies that foster this construction and growth.

Mobility of capital has played a crucial role in the growth of the United States. It has built many of the *Fortune* 500 corporations, skyscrapers, stadiums, and neighborhoods. It is now moving to other parts of the world on scales and with speeds greater than ever before. This motion is currently putting up luxury hotels in Beijing, golf courses in the Caribbean, and large numbers of factories in countries from Indonesia to Chile, served by airports and taxis. Some of these investments, like hospitals built by the World Bank in Tunisia, are seen by many as wise investments. Others of them are disputed by people in the countries that receive them. Like many economic stories, the quickening mobility of capital has a mixed history. One thing is clear about it, though—it is moving faster.

Susan George recounts the speed of capital mobility today with a description of the Mexican financial crisis. "In December 1994, billions of dollars were removed from Mexico in a matter of hours," she remembers. This was too much speed. Capital can build up quickly today in developing countries, but quickly means over the space of a few years, or at least a few months. With the sudden collapse of the peso, "interest rates were put sky high, over a million small businesses failed, unemployment became rampant, widespread hunger and malnutrition returned, crime rates soared alarmingly."[49]

Capital can move faster today than people can anticipate the effects of its movement. They might spend years planning and working to build a small business or a neighborhood association. But capital is traded without regard to the sudden changes that result from its mobility. On an ordinary day, this may not be a problem, because its speed may not be in excess and it may be well directed. But on a day of panic, or on a day when poor decisions are made by investors, the movement of capital can undo long-standing plans and social goals. A million transactions a minute now pulse through the New York Stock Exchange alone. In 1900 there were fewer than 2,000 trades per minute.[50]

These sudden changes can touch more people, in some cases, than were alive at any one time in the whole world before this century. The Asian financial crisis of 1998, when the value of the currencies and stock markets of countries from Malaysia and Indonesia to Japan lost value in a matter of days, likely affected more than 2 billion people directly, and others indirectly. But there were only about 1.6 billion people in the whole world at the turn of the twentieth century.[51]

Ownership in companies can be traded by the day or by the minute, without any long-term stake in the success and decisions of the company on the part of the person doing the trading. An individual can invest for his or her long-term future, and possibly even do so wisely, without simultaneously investing in the long-term health of the corporations whose stocks are involved. This disconnection reduces the imperative of companies to invest in their own long-term futures. Some important choices made by companies would be different if they were taken for the long term instead of for the short term. Policies of clean production and employees who are taken care of well are more likely to pay off over the long term, but may show up mostly as costs in the short term. But since individual investors' decisions are based on short-term results, and companies perceive their responsibilities to be first and foremost to shareholders, they are less likely to invest in the long-term health of the company.

Moving electronic capital creates physical motion. It moves shoes from factories to warehouses to shops; it sends trucks down the highway carrying new furniture and toys and other commercial items; it takes gasoline to gas pumps and metals to assembly lines. In these ways, it changes our physical world. When financial capital leaves agriculture and heads toward manufacturing, it reduces the use of one set of materials and natural resources while mobilizing a different set of materials. Fertilizer use can fall while hydrochloric acid use can rise. Moving financial capital from manufacturing to services can shift production to a whole different set of materials. These changes can result in a heavier burden on the environment, but they can also result in a lighter burden. They are not automatically for worse or for better. But they can transform economics, and nature, quickly and in ways that are unpredictable.

Moving electronic capital moves people as well. Changes in these capital flows have created large migrations of people from the countryside into cities around the world. From England during the industrial revolution to almost every developing country today, capital mobility is also human mobility. The speed at which the countryside of many countries is depopulated and the cities are filled is related to the trajectories and speeds that capital travels, whether it creates new jobs in the suburbs or the city, or whether it eliminates old jobs on farms. The developing countries have been undergoing a massive "rural-to-urban migration." It is driven in part by overpopulation and a lack of opportunities in rural, agricultural areas. But it is driven in part also by the generation of jobs in factories financed by new capital brought in from abroad or from a country's larger cities. The speed with which this new capital comes helps to determine the speed of those rural to urban migrations.

In China today, a "floating population" of as many as 160 million people drifts from countryside to city and then from one town to the next looking for work that is created, in part, by the shifting locations of capital, much of which is currently entering China from abroad. In the United States, these migrations took place more gradually, but they did take place, and they continue today.[52]

Large amounts of capital are moving quickly into slow places. The flow of international private capital, much of it from the United States, into developing countries, was only about $5 billion in 1970. In 1996, it had reached about $250 billion. It fell sharply in 1997, with the financial crisis in East Asia, but nevertheless remains high and is likely to regain its

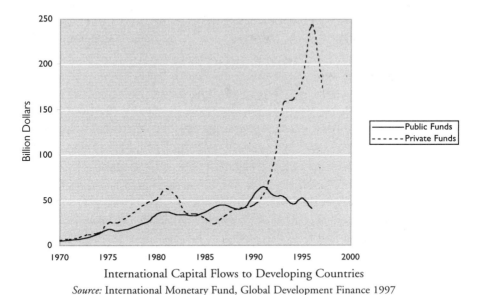

International Capital Flows to Developing Countries
Source: International Monetary Fund, Global Development Finance 1997

peak and rise above it during coming years. This motion of funds has brought industrial time to places that did not have it, from parts of rural China to isolated places in Malaysia, Laos, and throughout the world.[53]

Industrial time is kept by clocks and the rhythms of assembly lines in factories built just within the past few years in many places. Fast capital brings a unifying speed to all places of the world that it reaches, putting the production expectations of Chicago onto Phnom Penh and Xi'an. Industrial time does not always follow traditions. Its production schedules do not always have time for ancient holidays or sabbaticals. Industrial factories do not follow the seasons the way farmers do; they do not have to wait for daylight and can run right through the night if they have to; some of them operate during the siesta and on weekends.[54]

The speed of capital mobility makes farmers into factory workers and shoemakers into executives. It changes their working hours. It changes the amount of time they spend with their families. It changes the air they breathe, putting coal fumes into it instead of nitrogen fertilizer. It changes the water, which flows though cooling systems instead of trenches. The people who move this capital around the world do not make decisions of where to invest it based on the quality of the air or water, or the cultural changes that will accompany the new capital. But the decisions they do make form our world.

The Opportunity Cost of Taking Your Time

In 1953, two of Los Angeles's new expressways, one from Harbor to Pasadena and the other from Hollywood to Santa Ana, were linked together in a wholly new and original manner—the interchange. This concrete connector became a monument to nonstop traffic. On it, traffic regulated itself automatically. No traffic lights were needed. Not even a stop sign—at most, a yellow "Yield" sign would do the trick.[55]

Some thirty years later, on a day in 1984, near the small town of Caldwell, Idaho, a traffic light named "Red-eyed Pete," the last stoplight on the interstates, was removed, placed in a coffin, and ceremonially buried.[56]

We are gradually removing those institutions, like Red-eyed Pete and intersections without interchanges, that kept us slow. From traffic lights to poorly built roads, blocks to motion are seen as things to eliminate. We don't want to stop. A study at the National Zoo in Washington, D.C., found that the average time visitors spend looking at any individual exhibit is just five to ten seconds.[57]

Economists have popularized a concept that may explain our need to keep going, to go faster, to accomplish, and to make our economy grow. It is the "opportunity cost," the amount of money that we give up if we spend our time on any activity other than earning money. If your hourly wage is $20, and you take an hour off to go for a slow stroll around the neighborhood, then the stroll costs you nothing. But you still lose $20 that you would, otherwise, have earned. This clever concept puts an unfortunate burden on activities like the simple stroll that used to be free until the economics profession educated us.

For those who like to walk, or sleep, this concept makes poverty appealing. At the minimum wage of a little more than $5 an hour, taking a nap or playing with the dog costs $5 per hour. But for those unfortunate people who earn $50 per hour, taking a nap or playing with the dog costs $50 an hour. A night's sleep of eight hours costs $400 in lost income. A shower—$25 if you're not quick.

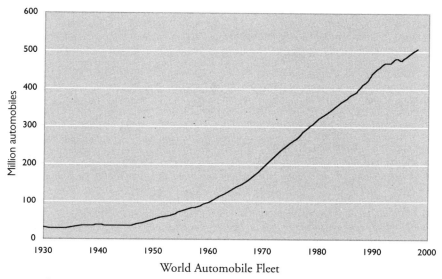

World Automobile Fleet

Sources: American Automobile Manufacturers Association, World Motor Vehicle Data; Standard & Poor's DRI; U.S. Bureau of the Census International Data Base, updated 30 November 1998.

This pressure may not serve human beings well. We evolved to need sleep, distraction, and guilt-free time off. Without these things, we become angry when the car ahead of us in traffic slows down. Without these things, people have higher rates of certain health problems. Without these things, people degrade their natural environment to an even greater degree.[58]

The greatest test of whether we can make our civilization into one that promotes our health, protects our environment, and meets our goals will be whether we can use time in ways that suit us. We are at a remarkable point in history, where the first billion human beings have obtained rapid communication and transportation, and where a second billion are about to join them. It is a point where many of us are working to eliminate old technologies and organizations that we now find to be too slow. It is a time when we can, and have to, shape a new way of life based on rapidity.

Ivan Illich has said that the opposite of fast does not have to be slow. It could be spontaneous. Or synchronized. "Quickening doesn't mean faster; it means coming alive," he says.[59] America's speed is transforming our lives and our environment. It is time for the country to come alive in more ways than just faster movement.

Travel slowed down by potholes or stoplights will no longer make these decisions for us. We will now have to decide ourselves how fast and how often we wish to travel, how much information we want to absorb, how frequently we want to move to new places, and how to make legion other decisions having to do with speed.

Notes

1. Miles per day figure from Alan AtKisson and Cliff Cobb, Redefining Progress, unpublished data, San Francisco, California.
2. Michael O'Malley, *Keeping Watch: A History of American Time* (Harmondsworth, Middlesex, England: Penguin Books, 1991).
3. "Matters of Scale," *World Watch Magazine,* November/December 1997.
4. Doris Flexner, *The Optimist's Guide to History* (New York: Avon Books, 1995).
5. Ibid.
6. AtKisson and Cobb, op. cit.
7. Bernd Polster and Phil Patton, *Highway: America's Endless Dream* (New York: Stewart, Tabori, & Chang, 1997.)
8. AtKisson and Cobb, op. cit.
9. Ibid.
10. Ibid.; Polster and Patton, op. cit.
11. AtKisson and Cobb, op. cit.; Polster and Patton, op. cit.
12. AtKisson and Cobb, op. cit.
13. Polster and Patton, op. cit.
14. Ibid.
15. Jay Walljasper, "The Speed Trap," *Utne Reader,* March–April 1997; Flexner, op. cit.
16. John E. Young, *Global Network: Computers in a Sustainable Society* (Washington, D.C.: Worldwatch Institute, 1993).
17. Walljasper, op. cit.
18. Ibid.; Flexner, op. cit.
19. Walljasper, op. cit.
20. Michael O'Malley, *Keeping Watch: A History of American Time* (Harmondsworth, Middlesex, England: Penguin Books, 1991).
21. Mark Kingwell, "Our High-Speed Chase to Nowhere," *Harper's Magazine,* May 1998.

22. Susan George, "The Fast Castes," *New Perspectives Quarterly,* Winter 1997.

23. Ibid.

24. Alan Durning, *How Much Is Enough? The Consumer Society and the Future of the Earth* (New York: W.W. Norton & Company, 1992).

25. Flexner, op. cit.

26. Ibid.

27. Young, op. cit.

28. Ibid.

29. *World Watch Magazine,* op. cit.

30. Wolfgang Sachs, "Wasting Time Is an Ecological Virtue," *New Perspectives Quarterly,* Winter 1997.

31. John DeCicco, James Cook, Dorene Bolze, Jan Beyea, *CO₂ Diet for a Greenhouse Planet: A Citizen's Guide for Slowing Global Warming* (New York: National Audubon Society, 1990).

32. Ibid.

33. Sachs, op. cit.

34. Ibid.

35. "Matters of Scale," *World Watch Magazine,* various issues.

36. Sachs, op. cit.

37. Calculations by Hal Kane, based on Stacy Davis and Melissa Morris, *Transportation Energy Data Book: Edition 12* (Oak Ridge, Tenn.: Oak Ridge National Laboratory, 1987).

38. Mary C. Holcomb et al., *Transportation Energy Data Book: Edition 9* Oak Ridge, Tenn.: Oak Ridge National Laboratory, 1987); Daniel B. Wood, "New 777 Loaded with Innovations," *Christian Science Monitor,* December 21, 1990.

39. United Nations, *1990 Energy Statistics Yearbook* (New York: 1992); Robert A. Egli, "Climate: Air-Traffic Emissions," *Environment,* November 1991.

40. Egli, op. cit.

41. Ibid.

42. William K. Stevens, "Global Warming Threat Found in Aircraft Fumes," *New York Times,* January 7, 1992.

43. Robert A. Egli, "Nitrogen Oxide Emissions from Air Traffic," *Technologie,* November 1990; Mark Barrett, *Aircraft Pollution, Environmental Impacts and Future Solutions,* World Wide Fund for Nature—UK/Earth Resources Research, London, August 1991.

44. Robert J. Samuelson, "The Buyer of Last Resort," *Newsweek,* August 24, 1998.

45. DeCicco et al., op. cit.

46. Sachs, op. cit.; Young, op. cit.

47. Young, op. cit.

48. Walter Russell Mead, "Trains, Planes, and Automobiles: The End of the Postmodern Moment," *World Policy Journal,* Winter 1995–96.

49. George, op. cit.

50. Kingwell, op. cit.

51. Calculations by Hal Kane, based on Population Reference Bureau, *World Population Data Sheet* (Washington, D.C.: PRB, various years); and PRB, unpublished data sheet of historical population figures (Washington, D.C., 1996).

52. World Bank, *China: Strategies for Reducing Poverty in the 1990s* (Washington, D.C.: World Bank, 1992).

53. Hilary F. French, *Investing in the Future: Harnessing Private Capital Flows for Environmentally Sustainable Development* (Washington, D.C.: Worldwatch Institute, 1998).

54. G. J. Whitrow, *Time in History: Views of Time from Prehistory to the Present Day* (Oxford: Oxford University Press, 1989).

55. Polster and Patton, op. cit.

56. Walljasper, op. cit.

57. Ibid.

58. Kenneth Jon Rose, *The Body in Time* (New York: John Wiley & Sons, Inc., 1988).

59. Ivan Illich, "From Fast to Quick," *New Perspectives Quarterly,* Winter 1997.

Casy said, "I been walkin' aroun' in the country. Ever'-body's askin' that. What we comin' to? Seems to me we don't never come to nothin'. Always on the way. Always goin' and goin'. Why don't folks think about that? They's movement now. People moving. We know why, an' we know how. Movin' 'cause they got to. That's why folks always move. Movin' 'cause they want somepin better'n what they got. An' that's the on'y way they'll ever git it. Wantin' it an' needin' it, they'll go out an' git it. . . ."

The moving, questing people were migrants now. Those families which had lived on a little piece of land, who had lived and died on forty acres, had eaten or starved on the produce of forty acres, had now the whole West to rove in. And they scampered about, looking for work; and the highways were streams of people, and the ditch banks were lines of people. Behind them more were coming. The great highways streamed with moving people.

—John Steinbeck, *The Grapes of Wrath,* 1939

CHAPTER TWO

Moving Away

*Americans Have a Unique Approach
to Accomplishing Their Goals*

Americans move to new houses and new cities more often than the people of almost any other country. Many of us climb the career ladder in a zigzag fashion through lateral moves to new employers. Some employers relocate entire industries to new regions. Half of our marriages end in divorce. We move away from our families at a young age to pursue new jobs and more education. We are shaping our future—through choices to move to new places, to new jobs, to different environments, and away from many of the things that we have known.

Moving—to a new region, apartment, career, social class—is as powerful a tool for social and economic change as politics. We could change our country through detailed dialogues about local politics. We could remake our neighborhoods to suit our needs by focusing inward instead of by leaving those places to look for new opportunities. We could remain in the same job through our lifetimes, continuing to use our skills, but we do not. Many of us move on to new opportunities and new places all the time.

Our army of realtors, abundant moving vans, inexpensive airplane flights, and legion of corporate personnel counselors and human resource specialists make it possible for us to leave our homes, jobs, and problems and search for new opportunities. Through most of history, rural farmers and craftspeople around the world had no choice but to remain where they lived and continue, year after year, with their work. But we have evolved ways of solving problems by using mobility and flexibility, rather than staunch dedication to one place, person, or job.

This is not to say that such mobility is a common ideal. Many of us cling to a historically ideal town that may never have existed, where passersby know each other, where shopkeepers know their customers, and where people stay in touch over their whole lives with their classmates from grammar school. But if that world ever existed, then it has since disappeared from many parts of our country. In the neighborhoods of most American cities today it is possible to walk down the main shopping street many days in a row and never see the same people twice. This is true from Washington, D.C., to Chicago, Seattle, and Santa Fe. The idea of a connection between passersby, or among diners at a restaurant, means little since they do not expect to see each other again. We do not have places of which we are a part; rather, we have business districts, airports, and highways that we pass through while we run our errands or do our business.

To move on is a rational act. It creates opportunities in new places, and escapes old frictions and discomforts. But it also weakens attachments. This changes the way that we treat our neighborhoods, our environment, and the future. We have less need to invest in and conserve the things and places that we expect to move away from eventually. We are less likely to contribute to local communities if we do not intend to stay in those communities. We invest less in the future of a place or an activity if we do not expect to share in that future. The phrase "planned obsolescence," created for objects like disposable watches, could even be taken to represent the ways that some of us think about some of the places where we live temporarily. When we move away from those communities, they become obsolete in our personal lives—and since we know that we will eventually move away, we have less reason to devote ourselves to those communities.

Moving Away from Home

An observer might think that Americans do not like their homes, given that they leave their homes at a remarkable rate. Some people spend their whole lives in their hometowns or in one house. But at mid–twentieth century, 21 percent of all Americans moved to a new residence every year.

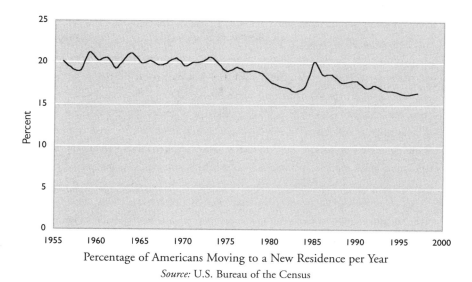

Percentage of Americans Moving to a New Residence per Year
Source: U.S. Bureau of the Census

Some people even moved more than once a year—and so were counted in this statistic more than once—while others moved less. But for every 100 Americans in 1955, twenty-one moves were made to a new home.[1]

All of the odds and ends that pile up in people's attics, and gradually become antiques over the years of dusty neglect, come into jeopardy when people move. Parents acquire the urge to throw out kids' baseball cards and dollhouses. Old rocking chairs go to the dump. Attachments to old elementary schools, to Little League, to local parks are forgotten when people unpack in a new city. Furniture is new, or newly acquired. And yet this activity is a national obsession. It is the remaking of the contents of our closets, even though the closets are part of where we keep memories and ties to past times and departed people. It is the discovery of new parks, with new playgrounds and new kids.

The rate at which we move to new homes has actually fallen during much of the second half of the twentieth century, due perhaps to the increased costs of housing in our ultrarich economy. Nevertheless, in each year of the early 1990s, 17 percent of Americans were moving to a new home (see figure).[2] Nearly 7 percent of those moved there from a different county, state, or country. During that period, the median length of time people ages 15 and over lived in their homes was 5.2 years.[3] In Tucson, Arizona, almost 30 percent of households have lived in their homes less than fifteen months. About 60 percent have lived there fewer than 5 years.[4]

Frequent moves are common not only to young people and those who rent, but to owners as well. Among people who rent, the median duration of residence is 2.1 years. People who own their homes, meanwhile, stay in them for a median length of 8.2 years. The duration of stay of many owners in America represents far less than a lifetime bond to a neighborhood. It also represents less than a long-term commitment to our surroundings.[5]

This residential train station that we call neighborhoods is more raucous in some regions than in others. Over the five-year period from 1985 to 1990, the U.S. state with the lowest mobility rate was West Virginia, where 36 percent of the population changed residence. The state with the highest mobility was Nevada, where 65 percent of all people moved to a new residence during that five-year period.[6]

The West is the most mobile region in the country, with a fifth of all households moving each year, followed by the South, where almost as many move.[7] Westerners move twice as often as Northeasterners.[8] In San Diego, Riverside, and other California cities, about one-quarter of all heads of a household moved to a new house each year during the early 1990s, a way of life that is continuing, for the most part. The same was true in Denver, Colorado. Over half the populations of most Western states moved between 1985 and 1990.[9]

The percent of U.S.-born population living in their state of birth runs from an extreme of 80 percent or more in Pennsylvania, Louisiana, and New York to less than 50 percent in states such as Arizona, Florida, New Hampshire, Colorado, and Oregon. More than three-quarters of the U.S.-born population lived in their state of birth in 1870 (about 77 percent). At the turn of the century, it had risen to about 78 percent. Even as late as 1940, three-quarters or more of Americans lived in the state where they were born. By 1990, the figure was about 67 percent—two-thirds.[10]

High mobility rates are not unique to the United States. Annual mobility rates of more than 17 percent are also found in Canada, Australia, Scotland, and New Zealand. But in France, Sweden, Great Britain, Switzerland, Israel, and Japan, annual mobility rates range from 9 to 15 percent. Other countries are more rooted still. The Netherlands, Austria, Belgium, and Ireland, for example, are among the industrial countries where fewer than 9 percent of the population moves to a new home each year.[11]

The departure from the places that we know has many consequences. Demographics professor Patricia Gober writes "Generations of high mobility . . . have depreciated the longstanding investments that people make in places—often called place ties."[12] In many ways, Americans don't know the places where they live. Ecologists complain that we don't know where our water comes from, or what kind of forests once covered the landscape, or which species were displaced by our construction. We do not know which tribes of native peoples were displaced or eliminated by our construction. Meanwhile, the close ties and personal knowledge that develop over decades-long relationships diminish with frequent relocations, and can end with too dramatic a displacement.

We have done this voluntarily. The government does not make us move on to the next place or experience. It is how we have chosen to live. In some other parts of the world, tens of millions of people are made to move every year to make way for new roads and power plants, to relieve overpopulation, or to escape hunger. Others flee poverty or war. Americans, though, relocate to advance careers, or just to try someplace new.

Leaving the Family

Young Americans waste no time today in moving away from their families. On average, women leave at age 19 plus a few months, and men leave today at close to 20 years of age.[13] This is often viewed as a good thing: separation from the nest makes young people independent, lets them devote themselves to colleges and schools, and may be important to their careers. It may even prepare them to withstand divorces later in their lives. This may be part of an American tradition of self-reliance—from an early age, we are an independent people practiced at moving out of the house and who can live and work in flexible ways.

Americans have always left home at a young age, though not always quite as young as today. In the early and mid-1930s, the median age at which men left home was over 21. It fell substantially during those years, and by the late 1930s it was only about 19. For women, it was just over 20 around 1930 and fell through the next two decades to about 19 at around midcentury. The age at which men left home began to rise again in the mid-1970s, almost reaching 20, where it has remained since.

In the late 1960s and early 1970s, women began to leave even a little bit earlier, younger than 19, for example, until an upturn in the mid-1970s.[14] Today's twenty-somethings, who came of age in the late 1980s, left at about age 19.5, later than any other cohort of young adults since World War II.[15] This upturn has rendered today's young people vulnerable to the accusation of having a hard time leaving home.

Rises in the ages at which young people marry are often cited as the leading cause for this difference, but difficulty in finding work, and other trends, also have an effect. Those who do not have income from a job to support them, or who do not have a spouse to support them, remain with their parents out of necessity, if not out of choice. Between 1960 and 1992, the median age at first marriage rose from 22.8 to 26.5 years for men and from 20.3 to 24.4 years for women. This rise in the age at marriage expanded the number of years in which young people are likely to remain with their parents.[16]

The description of Generation Xers as unable to move out of their homes says quite a bit about our expectations. It is an American expectation that young adults in their teens or early twenties should move out and devote themselves to education and a career. The proponents of economic growth must be delighted. Not only does this start people off toward productive working lives, it also means economic expenditures on moving vans, futons, and the materials used by people who move into dormitories or their own apartments. The proponents of strong family life, on the other hand, may not be so content. Coupled with a tendency for grandparents to have their own homes, or to live in retirement homes, the tendency for young people to move out means that parents live for many years with an "empty nest."

Free Agents

A textbook about management starts out: "The rate of leadership turnover is increasing, and the norm of organizational loyalty is weakening." Executives, and employees of many kinds today, do not hesitate to depart from their position or their company if a better opportunity arises. Some even believe that if they remain for long in any one job that they are "stagnating," and that their career is no longer progressing.[17]

The management book gives an example. "In 1979, Thomas Vanderslice was brought into GTE.... By 1983, two-thirds of the top ninety executives had been moved or newly recruited from outside.... Then Vanderslice departed abruptly when the chairman reneged on a promise to step down, and observers wondered if the new directions were sufficiently institutionalized to sustain the changes." If those four years were too short a time, then what is happening at all the companies whose leaders depart, willingly or unwillingly, after only one or two years?[18]

Earning a living now frequently involves leaving a job. Gone are the days of lifelong commitment to one trade or craft. Gone are the days of a marriage between a corporation and an individual. Getting ahead in the marketplace is about moving to those companies that are forging forward. Positions at firms with respected old names like General Electric or Boeing are tossed by the wayside if a better offer comes up, even from a riskier company that might not survive a few years of competition. In California, the median job tenure for workers is just three years.[19] During a typical year, about 14 percent of U.S. workers leave their jobs voluntarily. This is in addition to those who are fired or laid off or who otherwise leave involuntarily.[20]

At times, workers have drawn some of their identity from their employers. In Japan, this is still the case today for many people. Many Americans continue this identification as well. But many have abandoned it. They consult, temp, telecommute, work part-time, plan to leave, take "piece work," and follow many paths to their own goals rather than pledging long-standing commitment to an employer. The median job tenure for workers in California is three years, according to a Field Poll during September, 1998.[21]

The idea that a job does not last for a lifetime is a form of independence from the employer. It is even a form of independence from the livelihood. Many people today have enough savings that they can take some months off from work to rest, pursue new career training, or travel. Historically, in a nation with many farmers, no one took a season off from the

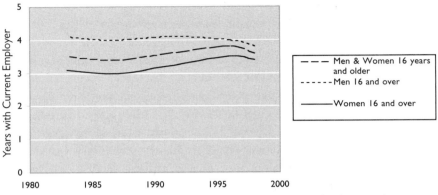

Median Years of Tenure with Current Employer, Wage and Salary Workers
Source: Bureau of Labor Statistics

harvest, because they needed its food; and few other people could walk away from their trades or from the other people who depended on their incomes. That flexibility is a recent phenomenon in middle class society.

People's identities were once so closely tied to their work that many surnames are derived from those trades: Smith, Shoemaker, Carpenter. This is also the case in other, non-English languages: Cohen (priest), Snider (tailor), and many others. But today we would never name ourselves for our work, in part because we change our work often and find our identity elsewhere. We have moved away from the constraints of one trade, and from a lifetime spent on one task. Political scientist Maryann Cusimano says that "identity is [now] just as mobile as the economy; you are not born with it, you can buy it." You can get it from the clothes you wear, the places you live, the hobbies that you pursue. Moreover, you can change those traits at will to change your identity.[22]

The loyalty between employers and employees is changing. Even in Japan, where many workers have, historically, remained with one employer through their whole careers, many are no longer staying as long. It marks a change in one of the fundamental relationships in society— employment. It is a change that severs traditional ties and commitments and, instead, makes us free to take new opportunities and make new choices.

This lack of devotion or loyalty in the workplace comes not only from employees: it also comes from—and may even originate with—employers. It is the industrial way. Over a typical twelve-month interval in the United States, about one in ten manufacturing jobs disappears, which means that 10 percent of workers can expect to be laid off each year, not counting those fired for other reasons. Meanwhile, a comparable number of new manufacturing jobs also open up—but at different locations and different companies and using different skills. Some of these jobs even return later to their original sites, companies, or individuals. Most of them do not reopen at the same location within the following two years, however. Workers have to shift, adapt, move, and wait.[23]

The industrial economy shifts work to suit the demand for new products. Its loyalties are to efficiency and consumer demand. The critical ties are not those between employer and employee but rather those between producer and consumer. Jobs are moved to serve that other primary relationship—they are shifted from one geographical location to another depending on where the production facilities are located for a product that is selling well; they are eliminated if demand falls for the product that they make; people are retrained to do a new kind of work or make a different product if the marketplace calls for that new product. It is not a marriage between people and organizations; it is a love affair between the seller and the shopper. One of the results of this relationship is an abundance of affordable commercial products and an economy coveted by people around the world. Another of the results is a severing of the long-term relationship between employer and employee, because those employees will be fired or retrained to suit fluctuations in consumer demand.

Jobs are created and destroyed as fast, or sometimes even faster, in other countries. In most European and industrial countries, and in some developing ones, old jobs are eliminated and new jobs are created, for new people in new locations, as frequently as in the United States.[24] Among European countries in recent times, only Norway and Germany have had less elimination of old jobs and corresponding creation of new ones. The economic system that we use to provide us with large, warm houses is one that has relocation and movement built into it inherently, because it requires workers to move to new kinds of work, to move to new locations, and to develop new skills frequently.

The very idea of a "job" is a relatively recent one, originating in Victorian England. Work has existed through all of time—but it consisted of tasks, often endless tasks, that a person or a family performed to feed themselves, shelter themselves from the weather, and meet all of their needs. They did not "possess" a job. Rather, they had work to do, either for themselves or for a feudal ruler or other political governor. Nevertheless, the job has become a fundamental component of our culture. It is where we go when we get up in the morning. It is where we learn many of our skills and meet many of our acquaintances. Many people spend more time at their jobs than on any other activity.

Social critic Walter Russell Mead traces separation from the family back to the time in Victorian England when people's work became their "job." Victorian capitalism, he says, was the first to have large autonomous factories and office sites, where workers were separated from home when they worked. Families no longer worked together very frequently as units, he says, the way farming families had worked together. Blacksmiths' shops were often part of their homes, and their families were nearby, sometimes even as helpers. Small farmers worked on their own land or on the land that they lived on, often with their families by their sides. Craftspeople worked at home. But the modern concept of a job involves a separation between home and office, or between family and factory. It came about as part of increasing specialization and the greater efficiency that specialization allowed in production. Office workers specialized in paperwork, meetings, legal proceedings, and so forth, while assembly line workers went to factories, and builders went to construction sites. All of these people left their homes and families to go to those locations. Today's downtown office is a historical anomaly as a place separate from the home and the family.[25]

As some Americans abandon their secure jobs, they are returning to a historical form of work, by working at home or in other, alternative ways. This return can actually reestablish some lost connections. Before the historical time of the "job," work and private life were more mixed. Returning to a trend of work instead of a job can mean reuniting work, family, and home. Rates of working at home are on the rise today.[26]

The departure from the office, to the extent that it is happening, is not at odds with economic goals. It is a form of flexibility that allows companies to hire people for the short term and then not be burdened with them later. It allows workers to have freedom from long-term employment. It adds flexibility to the economic system. It is happening along with a rise in entrepreneurship and the creation of new, small companies. Silicon Valley, for example, has an abundance of these flexible working relationships, of workers working unusual hours, of people telecommuting, and of consultants and temps.

Some people have accused firms of trying to avoid paying benefits and of not standing by their workers. In some cases, this appears to have been true. In many cases, though, it is the workers who have created these flexible relationships. Many other people, however, view these flexible and temporary work relationships as a virtue. Many workers have taken advantage of these flexible relationships to become consultants and entrepreneurs, and to launch organizations based on ideas for new products or services. Many of Silicon Valley's own Internet start-up companies and consulting firms started because their founders had the flexibility to work as consultants or to become entrepreneurs. Other people have gained the ability to work from home and, as a result, can be near their children and families while still being able to earn a living. Still others have become able to design their own unconventional schedules to allow themselves to pursue their hobbies or other interests during daytime hours and then work at night or on weekends.

Some Japanese economic analysts have attributed a lack of entrepreneurship and venture capital in Japan to the more rigid lifetime-employment system there. Japan's microcomputer industry, they say, is immature because too many people stay at large companies rather than splitting off to form new, agile start-ups. They say that even though the lifetime-employment system is starting to crumble, large companies still manage to retain their best and brightest with the lure of job security and high pay. "In doing so, they discourage job mobility—and creativity, too, since good ideas rarely make it through large corporations' bureaucratic structures," wrote Irene Kunii in *Business Week* magazine.[27]

This means that the long-standing ideal of secure employment is diminishing, even in places like Japan, according to wishes of people from backgrounds as diverse as economic strategists, workers, and executives. This will pry many people free of places where they have stayed for decades. It will mean the gain of their skills to new employers and the loss of their skills to old employers. It is a loosening of economic ties. It may be one of the most important economic developments of our day. It is also a social loosening. It is one of the most important social developments of our time.

Wandering Industry

Our industries have become more footloose, disconnected from their geographical locations. The products they make are often not related to the places where the company is based. Many offices are located in one region, while the factories that produce their wares are in another region or another country. Consumers are distant as well, not needing to live anywhere near the place where their purchases are made. "Within the logic of commerce, location is an increasingly meaningless concept. We confront, now, a new topology, a world of instant and direct contact between every point on the globe," wrote Mark Kingwell in *Harper's*.[28]

The world's largest wholesale distributor of computers, for example, is called Ingram Micro, Inc., and it assembles computers not for itself but for several of the world's leading manufacturers. Ingram Micro has a conveyor belt three-quarters of a mile long in a converted warehouse, and produces computers for Compaq, IBM, Hewlett-Packard, Apple, and Acer. Ingram Micro packs them in boxes that carry not the Ingram logo but the logos of their clients, and ships the computers directly to the customer with a label that makes the box look as though it came from a local dealer. This company is based in Santa Ana, California, but that does not matter. No connection is necessary between buyers and the company that they are buying from.[29]

Ingram Micro even installs software directly onto the computers, sets up World Wide Web sites for far away clients, answers phones in their name, and turns bill collectors loose on their deadbeat customers. All of this almost completely absolves its client companies from working with their customers, and makes geographical location irrelevant. The *New York Times* called this "a glimpse of the future of American industry, where manufacturers don't make anything and retailers don't touch the goods they sell." In this new world, the industries can easily locate themselves far from the places where their customers, natural resources, or other concerns are located.[30]

Today, most of the international work done by corporations is done through trade with firms in other countries, and by foreign ownership of stock, rather than by relocating or building new companies in foreign countries.[31] A million transactions a minute now pulse through the New York Stock Exchange (NYSE) alone—whereas in 1900 there were fewer than 2,000 such transactions per minute. This investment in the stocks traded on the NYSE makes money and equipment available for activities that would not be possible otherwise, in places that would not be involved otherwise.

As a result, decisions made by investors in one location affect people and nature in far away places. Investments held in Manhattan make it possible for companies to log forests in Cambodia. Boardrooms in Los Angeles are the site of choices that affect people and nature in Mexico and Argentina. But those transactions are anonymous. Most investors will not see the places where the companies that they own have facilities for mining, logging, fishing, shipping, or whatever work they do. Our business decisions have, over the years, moved increasingly away from the places that they affect.[32]

Significant amounts of international work are also accomplished through corporate relocations, and to those people whose employers ask them to move, it is a life-changing shift. Patricia Gober says that "IBM" stands for "I've Been Moved."[33] The phrase singles out IBM because its acronym contains letters that can be used to illustrate the trend in relocations; most other large companies could have been chosen instead if their names had leant themselves to the pun. Corporate relocations are a normal business practice, consistent with the management practices taught in business schools and respected by top managers. Businesses that relocate do not have reputations for being "fly by night"; they are respectable corporate citizens. They are not long-standing members of a community, however.

Many people, from politicians of the far right to those of the far left, to union workers at automobile factories and the makers of computer chips, have debated the harms and merits of "globalization." What they mean by this term is increases in international trade, international investment, and frequent relocations of companies, as well as an increasingly international civilization. Most literature about globalization says that trade and foreign investment have reached unprecedented levels and threaten American jobs and ecology.

In two respects, however, the world was as globalized in 1913, just before World War I, as it is today. The world's levels of international trade and foreign direct investment at that time were at about the same levels that they reached at the end of the twentieth century. They are measured by the dollar value of the merchandise and services that are traded across international borders, and the dollar value of investments made in foreign countries. The annual worldwide rate of growth, in dollars, of merchandise trade was 1.7 percent a year from 1870 to 1913. But two world wars and the depression of the 1930s knocked economic integration down to far smaller levels. During the years since 1950, the merchandise trade has grown again, at 1.3 percent per year. Foreign direct investment financed 9 percent of the world's total production in 1913. In 1960, it was just 4.4 percent. By the mid-1990s it had risen to 10.1 percent, finally passing the level of 1913.[34]

Here is the difference. In the world of 1913, globalization meant increasing trade and investment. It even meant migrations of workers, although not in numbers as large as today. But, for the most part, it did not mean the relocation of industries to foreign countries. Industries were more likely to move inside their own country than to leave it. In the world of the year 2000 current globalization takes place not only through trade and investment and migration but also through the uprooting of industries from their historical locations. This may be what is most threatening to many constituencies and individuals. It is not the trade of medical technology to fight illness abroad. It is not foreign ownership of U.S. skyscrapers or icons. It is the fluidity of location of industry and job location that threatens the livelihoods of many.[35]

In the technical language of economic literature, this motion of industries is called "capital mobility." The fundamental economic theory often invoked to claim that international trade is beneficial to all countries involved is Ricardo's theory of "comparative advantage." It states that international trade allows countries to specialize in those products that they can produce the most efficiently, thereby improving their economic performance; they can then trade to get the other products that they need but are unable to produce efficiently themselves.

But this theory explicitly states that capital mobility must be low in order for trade to spread benefits among all countries—otherwise the advantages of specialization can be lost to overall dominance by wealthier countries endowed with larger amounts of capital and ability to produce. It says that trade benefits all countries involved, and that countries should specialize in certain goods, provided that the trade in merchandise takes place at a time of "low capital mobility." But this theory is applied falsely if it is applied at a time when corporations and industries, and capital, can relocate easily to new countries or new regions.[36]

If an underlying assumption of the theory does not hold then the theory does not apply. Many university economics departments teach that "trade is good," according to strict economic theory. But they do not teach that one of the basic assumptions of this theory does not apply. This basic assumption is that capital is not mobile, that industry will stay home. This may be the real source of tension today that explains why globalization is debated angrily in political forums. The time when globalization becomes threatening is when the movement of merchandise, investment, migrants, and whole industries themselves all function together. It is a time when people are not only separated from industries, but when places, people, and commerce all flow freely in many directions, threatening our senses of stability and control.

Civic Disengagement

One of the most powerful effects of our moves to new places and new kinds of work is that they separate people from each other. When we move, we leave the local organizations in which we had previously been members. We leave the relationships that we have built up over months or years of interaction. Harvard professor Robert Putnam believes that civic associations in the United States are in decline. He has gathered evidence from Americans' membership in various kinds of organizations, and he paints a picture of a more private and isolated country in which people have moved away from each other.

Putnam's evidence begins with voter turnout, which declined by nearly one-quarter between the early 1960s and 1990. Next, he points to a modest decline in churchgoing. From about 48 percent of Americans in the late 1950s attendance at church fell to about 41 percent in the early 1970s, which is about where the rate is today. Membership in labor unions fell steeply as well. Since the mid-1950s, when union membership peaked, the unionized portion of nonagricultural workers has dropped by more than half. From 32.5 percent in 1953, union membership of these workers was 15.8 percent in 1992. "By now, virtually all of the explosive growth in union membership that was associated with the New Deal has been erased. The solidarity of union halls is now mostly a fading memory of aging men," he says.[37]

Putnam's evidence continues. The parent-teacher association (PTA) has lost members. Membership fell from more than 12 million in 1964 to barely 5 million in 1982, before recovering to about 7 million in the mid-1990s. Membership in women's organizations, Putnam says, has also declined more or less steadily since the mid-1960s. Membership in the national Federation of Women's Clubs is down by 59 percent since 1964, and membership in the League of Women Voters has dropped by 42 percent since 1969. The Boy Scouts are down by 26 percent since 1970; the Red Cross is down by 61 percent since 1970. "Serious volunteering declined by roughly one-sixth" from 1974 to 1989 says Putnam. Meanwhile, Lions Club membership is down 12 percent since 1983; the Elks are down 18 percent since 1979; the Shriners are down 27 percent since 1979; the Jaycees are down 44 percent since 1979; and the Masons' membership is down 39 percent since 1959.[38]

This barrage of falling memberships is convincing evidence of a change in American gathering habits. They suggest that we are disengaging from the civic organizations of our past, and that we are moving away from our traditional gathering places. These changes are as fundamental as most of the topics that fill our newspapers. It is tempting to think that the only reason why such reductions in organizational membership and participation have not been discussed more often in forums of public affairs is that those forums have lost their members to the point where they no longer have the ability to direct attention to the loss.

This does not mean that we are becoming less active. On the contrary, data show that many people are working longer hours, traveling frequently, spending more time on the Internet. But falling memberships are not an indicator of less activity—they indicate that we are associating less with each other. We do much of our work by ourselves or with a small number of colleagues; we spend time with technologies like our computers or our TVs. These new forms of entertainment and long hours of work are replacing some civic forums.

It has been said that America's greatest resource is its people. If its people are disassociating from each other, then they are giving up some of the interactions, networks of individuals, and community institutions that are our greatest resource. If we learn more skills individually, enhancing what has been called our "human capital," then this will still not compensate for reductions in the social and civic interactions that have been called our "social capital." Better computers will not substitute either. More roads will be no consolation. We have gained freedom in the move away from civic organizations—we no longer have to be on time to as many civic meetings, we no longer have to follow organizational rules and procedures. But we have given up membership, which is one of the most fundamental human needs.

Putnam's signature piece of evidence is that even though the number of bowlers in America increased by 10 percent between 1980 and 1993, participation in league bowling fell by 40 percent. "Lest this be thought a wholly trivial example," he says, "I should note that nearly 80 million Americans went bowling at least once during 1993, nearly a third more than voted in the 1994 congressional elections and roughly the same number as claim to attend church regularly."[39]

A question asked by Putnam and many who have critiqued his study is whether membership in other organizations grew during those years, offsetting the declines. Putnam points to organizations like the Sierra Club and National Organization for Women, which grew rapidly. The American Association of Retired Persons grew exponentially from 400,000 members in 1960 to 33 million in 1993 to become the largest private organization in the world after the Catholic Church. But Putnam points out that for most of these members,

> the only act of their membership consists in writing a check for dues or perhaps occasionally reading a newsletter. Few ever attend any meetings of such organizations, and most are unlikely ever (knowingly) to

encounter any other member. The bond between any two members of the Sierra Club is less like the bond between any two members of a gardening club and more like the bond between any two Red Sox fans (or perhaps any two devoted Honda owners). . . . Their ties, in short, are to common symbols, common leaders, and perhaps common ideals, but not to one another.[40]

Nicholas Lemann, however, does find some possible examples of organizations that have grown in counterbalance to those that have declined. U.S. Youth Soccer, he says, has 2.4 million members, up from 1.2 million ten years ago and from 127,000 twenty years ago. "As a long-standing coach in this organization, I can attest that it involves incessant meetings, phone calls, and activities of a kind that create links between people which ramify, in the manner described by Putnam, into other areas," he says.[41]

Lemann also offers the rising number of restaurants in the United States, which has risen from 203,000 in 1972 to 368,000 in 1993, as evidence for new associations among Americans. And he points out that individual contributions to charity, which are still made by more than three-quarters of Americans, grew from $16.2 billion in 1970 to $101.8 billion in 1990. But these restaurants and contributions fall into the category of common ideals and consumptions, not community among people. Writing a check to pay a restaurant bill or to help a charity means giving money rather than giving one's own time and direct interaction. They are a substitution for membership—a moving away from direct contact.[42]

If Putnam is right, then America is a more private country than it was in the past. This privacy means much for politics, and for the future. If our chief gathering places are offices, where we produce commercial products and earn money for our services, then our civic existence will largely be an economic one. Other values, practiced at voluntary organizations and social clubs, will take a backseat if our social interactions are more often centered on jobs. With fewer extrafamilial social institutions, we spend less time thinking about the needs of the people with whom we formerly shared an identity as members. And if our chief remaining social unit is the family, and kids move out of home at nineteen and a half years of age, then even that social unit is less robust than it once was.

Politics will be affected in many ways by changing tendencies to associate. Members of associations are much more likely than nonmembers to participate in politics, to spend time with neighbors, and to express social trust. A 1991 survey called the World Values Survey demonstrated that across thirty-five countries surveyed, social trust and civic engagement were strongly correlated: the greater the density of associational membership in a society, the more trusting its citizens.[43]

If the energy that we spend moving to new homes, looking for new jobs, relocating entire industries, and shopping for new disposable products went toward local politics instead, then we might have politicians who looked farther into the future than the next election. If that energy went toward local business, then we might have industries that made longer-term investments in their local surroundings. Such a concentrated, local commitment might become a foundation for quality local newspapers and meaningful local civic debates. But that is not where most of us put our energy.

Moving Away from the Poor

In a country that prides itself on equality, the rich are moving away from the poor in several ways. The first way is through their rising incomes. Differing incomes, in turn, often lead people to move apart to separate living quarters, separate parts of town, different career opportunities, and different memberships in clubs and other affiliations. This split begins with income, but then it changes the country.

If the rich get richer at no expense to the poor then that might not represent a loss to the country. For years, that is what happened. John Cassidy wrote in the *New Yorker* that up until about 1973, poor and rich Americans were all getting wealthier. Families that earned $10,000 a year in 1947 got about $20,000 in 1973. Those with $25,000 in 1947 got $50,000 in 1973, he says. Those with $50,000 got $100,000—a doubling of real income for each group.[44]

But that soon changed. By the end of the 1970s, an American who earned the national median income, in constant, inflation-adjusted dollars, was earning $498 a week, or $25,896 a year. In the middle of the 1990s, the same person was earning $475 a week, or $24,700 a year. In sixteen years, he or she had suffered a wage cut of about a hundred dollars a month, or 4.6 percent, according to Cassidy. The median worker had taken a step backwards in income.[45]

Meanwhile, the wealthier segments of the country became wealthier. A full-time worker in the top third of the country's income distribution earned $890 a week in 1979—which comes to $46,280 a year. By 1995, he or she earned $960 a week—$49,920 a year—an increase of 7.9 percent. The decline of income for the median worker contrasts with this significant rise for the wealthy.[46]

The richest 5 percent of American families, however, earned $137,482 a year in 1979 on average. By 1993, that amount was up to $177,518—an increase of $770 a week, or 29.1 percent. The top 1 percent of families, meanwhile, did far better yet. Twenty years ago, the typical chief executive officer of a large American company earned about forty times as much as a typical worker did, and now earns a hundred and ninety times as much, according to Graef Crystal, an expert on executive compensation.[47]

Between 1977 and 1999, the income of the lowest one-fifth of earners fell by 12 percent after adjusting for inflation. The next lowest fifth of American incomes fell by 9.5 percent during that period. The middle fifth saw their incomes fall by 3.1 percent. By contrast, the next highest group had their incomes increase by 5.9 percent. And the top fifth enjoyed a rise in income of 38.2 percent. The highest 1 percent enjoyed an increase of 119.7 percent, adjusted for inflation.[48]

The growing separation between rich and poor is physical as well as financial. Walter Russell Mead writes that long ago, in Victorian England, "rich and poor mingled together"; but that the invention of the railroad separated classes in a socially stratified geography with the poor living in crowded areas near factories and the wealthy traveling quickly away from those places to expensive homes. He writes of "[t]he dark tenements of the working classes—in which men, women, and children were all condemned to incessant toil—were found in and near the polluted factory districts where production was king," while other people moved to suburbs.[49]

The train, combined with a historical split in wealth, became a tool of moving the rich and poor to separate quarters. In the United States, as early as 1832 the New York and Harlem Railroad offered a commuter rail

Income Disparity in the United States, 1977 and 1999

Household Groups	Average After-Tax Income		Share of All Income (est. percentage)		Percentage Change
	1977	1999	1977	1999	
One-fifth with lowest income	$10,000	$8,800	5.7	4.2	−12.0
Next lowest one-fifth	22,100	20,000	11.5	9.7	−9.5
Middle one-fifth	32,400	31,400	16.4	14.7	−3.1
Next highest one-fifth	42,600	45,100	22.8	21.3	5.9
One-fifth with highest income	74,000	102,300	44.2	50.4	38.2
One percent with highest income	234,700	515,600	7.3	12.9	119.7

Sources: David Cay Johnston, "Gap Between Rich and Poor Found Substantially Wider," *New York Times,* 5 September 1999; Congressional Budget Office data analyzed by Center on Budget and Policy Priorities.

service. This allowed people who could pay for tickets and for real estate to live apart from their jobs and apart from crowded places. Historian Clay McShane says that Boston probably had more than 2,000 daily rail commuters by 1850. He says that railroads saw those daily riders as a profitable short-haul business and encouraged their ridership by giving them low monthly rates, "a practice known as 'commuting' part of the fare." In Massachusetts, the state government encouraged the trend by mandating rebates for commuters because it believed that this would relieve urban housing congestion.[50]

In some cases, it was not the rich, but rather the poor who moved out to the suburbs, away from places where desirable central apartments fetched high prices. In all cases, though, fast transportation was shaping the distribution of wealthy and poor, usually as separate from each other. Since those early years, mobility has continued to play a role in social stratification. Wealthier people have fled urban blight over the years, to the point where many of these people no longer return even for a few hours to the urban centers that they or their parents once left. And their flight, in turn, contributed to the blight by removing some of the wealth and the human and social capital that was once in those urban areas. The image of these people moving to gated communities in the suburbs is matched by the image of their wealth, skills, and business networks also being locked behind closed gates.

Henry David Thoreau was a suburbanite who took advantage of the train to separate himself from the city while not losing its occasional benefits. He picked Walden Pond, a site half an hour from Boston by train, in part because it was near a railroad. He built his hut less than 200 yards from the railroad and within a short walk of the suburban homes of his commuting literary friends in Concord.[51] Today, though, he could go even farther away, and could use a car to free himself from the train tracks as well.

Commuting used to be a phenomenon that went from suburb to city center. But now it is becoming increasingly a pattern of motion among suburbs and, in the case of many people, never requires a trip to the city center at all. Private cars allow suburb-to-suburb trips along routes that no bus or train travels. High personal mobility makes all of this possible, and allows people to dissolve the historical places and commuting lines that were once crucial to how a city and a region functioned.

The car has been instrumental in redefining the metropolitan landscape. The urban geography of the freeway city, which branches off in as many different directions as there are roads, is much harder to read than the linear one of the train, says Mead. People who travel by train have only as many options as there are train lines. But the car allows people to move quickly and easily in the directions offered by all of the streets and highways that fill our cities. According to Mead, "the railroad imposed mass schedules and transport patterns on large groups of people. The automobile dissolved them. Railroad society was table d'hôte; the Automobile Age was a buffet. Passengers went where the rail systems took them; drivers went where they pleased along the paths laid out for them."[52]

More than the trains did, the car has allowed some areas to be "gentrified," as wealthy, often white, people buy up urban neighborhoods formerly filled by a different ethnic group. Examples are abundant. The entire city of San Francisco, many people believe, will become gentrified now that its average housing prices have passed those of Manhattan. Many San Franciscans work in Silicon Valley, sixty or more miles away, and drive from the city down the highway to work. The availability of the car is shaping the distribution of wealth in San Francisco and putting computer workers and executives in place of earlier inhabitants. A city magazine there called the *Guardian* has labeled this the "economic cleansing" of San Francisco.

The separation between rich and poor enters politics in many ways. Poorer areas are less able to lobby for political benefits. State and national politicians themselves often come from wealthy areas and wealthy families far more often than from poor ones, and many of them are independently wealthy long before they are elected to Washington, Sacramento, or Lansing. Police forces are better funded where the wealthy live. Waste dumps are located in poor areas more often than in rich ones. These discrepancies build a cycle where the rich move ever farther away from the poor.

Some of the squalid conditions in slums and rural places exist because of the separation between those who own buildings and factories and those who live and work in them. The image of the slum landlord who makes money off of a decrepit building while living in luxury is based on the geographical dislocation of the two. If the owner lived on site in the rental building, then it would be less likely to be decrepit. The owner of an enterprise is less likely to take advantage of familiar people than of those who are strangers. It is easier to take advantage of people who are out of sight than of those who are seen.

This division is part of the American economic system. Stockholders live away from the companies that they own; mutual fund owners often do not even know what companies they hold. People from poorer neighborhoods who begin to earn large salaries often leave their old neighborhoods. Their new wealth becomes a division between them and the place where they grew up or where their family and friends live. Their new skills are lost to that place. It is a micro-scale version of the "brain drain" and departure of wealth that plagues many third world countries. This dislocation causes problems for inner-city and rural areas that find themselves in the quicksand of poverty.

This anonymity of ownership makes responsibility less crucial and makes irresponsibility less embarrassing. The geographical places where people who own large amounts of stock live are not the places where most manual labor is done. They are not the places with the worst environmental problems. This separation causes a mismatch of problems and financial resources. If our wealth were located in the most troubled areas, then those troubles would be more likely to be addressed.

This is as true of America as it is of the world as a whole, of course. American and European wealth is separated from much of the rest of the world. There is a divide between the developed and developing worlds that is, in some cases, larger than the divide between Westchester County and the Bronx, or the northern and southern suburbs of Chicago. Measured in dollars, even the poor of the United States are often wealthy by the standards of South and Southeast Asia, Latin America, Africa, and other places. Chaos and hunger taking place in some of those regions shows up only in our newspapers, or is not noticed at all by Americans.

If this wealthy country had to face those problems directly, then maybe it could, and would, turn its knowledge, technology, and money to trying to solve them. If our national borders did not separate our relatively wealthy population from poverty in Mexico, Rwanda, and Bangladesh, then our expertise would flow more often to those places. The dislocation between rich and poor is a device that we use to help us accept poverty. We accept it by distancing ourselves from it—and, lately, by moving even farther away from it.

Moving Indoors

Human beings lived through most of history at the mercy of nature. They were vulnerable to the cycles of the crops, the cold, insects and animals, and all environmental fluctuations. We have worked hard to try to escape this vulnerability. Some historians trace our first success in this effort to Victorian England. The citizens of that time were the first human beings, they say, to live lives in which natural forces did not determine their diets, their work, and even their lifespans. Instead, flows of financial capital determined their livelihoods, at least in part.

Walter Russell Mead says "previous generations had their fat and thin years, but these depended on the cycles of crops and weather rather than on the mysterious fluctuations of the stock and commodity markets. This was the first era in which human prosperity depended visibly on economic rather than on natural cycles."[53] The generations that have followed built upon this Victorian separation from natural fluctuations, almost to the point where many people do not see, touch, or smell nature during their daily lives. Our food comes from the supermarket; our water comes from reservoirs that can be hundreds of miles away. Some of us work at night and sleep in the day. We can wear shoes for months and yet never get dirt on the soles because we walk entirely on pavement and carpet.

In many places, we have separated ourselves from nature so much that our buildings are designed in ways that no longer fit their surroundings. James Howard Kunstler says that when we build, we build "tract developments in far-flung suburbs and isolated split-levels in cornfields, pastures and woods. They will relate poorly to things around them, eat up more countryside and increase the public's fiscal burden. . . . Our building practices are wholly at odds with our notions of what makes a place worth caring about."[54]

Travel writer and historian Robert Kaplan says that "our 'civilization' has disconnected itself from the environment around it. Homes in Arizona are not built in adobe anymore. And they have lawns—front lawns, full of short grass, watered from aquifers far beneath the ground that only refill slowly over hundreds of years." These are among our moves away from nature; and they not only change our experience, they also affect nature itself deeply.[55]

Under historical water doctrines in many western states, every drop of water can be claimed from the rivers and streams, with none left to flow to the ocean or provide habitats to wildlife or play a role in ecology.[56] The legal system did not even acknowledge the existence or the value of ecological health. Water is only a commodity to use, not a part of nature. Our code of conduct, the one that protects us from crime and violence and that regulates our commerce and other activities, puts our actions directly at odds with the health of rivers and streams. Through this legal system, we conceive of ourselves as independent from the environment. "A place that receives less than about twenty inches of rain a year has difficulty sustaining a large human population. But Tucson only gets eleven inches, and Phoenix and El Paso only get about eight inches of rain a year. Their aquifers are depleting," says Kaplan.[57]

Consumers would never know that this is going on (as long as it doesn't make them sneeze). Their experience is disconnected from water and other environmental resources to the point where they only experience the products they buy, not the ecosystems that provide them. The supermarket separates people from the toil involved in growing food, from most of the pesticides and fertilizers used to make it grow fast, from the smells of the fields, from the sun beating down on their necks, and from any possible injuries or discomfort during the harvest.

We can sit in air-conditioned cars, listening to music, isolated from the pavement by fabric and rubber, and isolated from the dirt by pavement. On either side of the road, landscaped vegetation sits in orderly rows and quadrants. The suburbs sprawl on into the distance, and sky-rise office buildings are surrounded by parking lots. Nonprofit organizations have to create special trips designed to take inner-city kids out to the countryside so that they can actually see it. Without such programs, some of these children would never see mountains, the ocean, forests, or lakes. The experiences of farmers and explorers would mean nothing to them. Yet we will depend on those children to make rational decisions about environmental management, commerce, and government in the future.

We depend on them already, since many of the kids who grew up all through the 1960s, 1970s, and 1980s had little contact with nature. Entertainment had gone indoors by the time they were growing up. Rentals and sales of videocassettes overwhelm hiking trips and evenings in the backyard. The rising use of VCRs is a movement indoors, a substitution of the home box office for a trip across town and of film for nature. It is the movement indoors, away from the environment.

Some people believe that the environment will never be healthy under human management if people's experience is separated from it. Wendell Berry writes, "Land cannot be properly cared for by people who do not know it intimately, who do not know how to care for it, who are not strongly motivated to care for it." He continues "People are motivated to care for land to the extent that their interest in it is direct, dependable, and permanent."[58]

These conditions often do not hold in the United States today. Most people do not have a dependable interest in maintaining the health of the land. They do not think that it matters to their incomes, their careers, or their homes, at least not in a direct sense or in the short run. People do not have a sense of permanent attachment to the land that they live on or the land that they eat from. After all, they may leave it soon. Berry's "direct, dependable, and permanent" interest vanishes when people move to new houses frequently and leave the places and people they know.

There is a connection between these changing human values and the erosion, logging, mining, overharvesting, waterlogging, and other harms to the land that are taking place today. One of the reasons why water tables are falling in many places is that the natural ecology of water is neither accounted for in our economy nor sensible to our daily household experience. The water has great value to us, but we often do not recognize that value because it comes out of our taps and showers at the turn of a handle. We consume large amounts of water when we eat vegetables, grains, and meat; but we only recognize that consumption as being about the foods we see in front of us rather than the water that was used to produce them. When we think of our utility bills, we often think of electricity, gas, telephone service, and garbage collection before water.

The Colorado River no longer reaches the ocean during some parts of the year, and American water obligations to Mexico go unfulfilled as that river's water is consumed upstream. This situation is at odds with our traditional values of responsibility to our neighbors, whether they be individuals or countries, and of environmental protection. Yet we violate those values when it comes to water because we rarely experience the river and rarely talk to the Mexicans who also need it. Our daily experience is as isolated from the Colorado River as it is isolated from the Mexicans who are separated from us by a border of barricades and arms.

Likewise, the natural ecology of soils, forests, the atmosphere, and many other aspects of the environment make no sense to our economy or daily experience. Climate change is an issue today in part because we cannot conceive that the act of driving a car could change the weather. Pollution exists in part because we do not view industry as something that could cause emphysema in our grandparents. Species go extinct, even the big, fuzzy, cute ones that people like, because in our minds our activities are separate from those species.

Even the prices that we pay for our material purchases are separated from the environment. Government subsidies and other determinants of price do not reflect the values, realities, costs, or experiences of nature. Water costs far less when priced in dollars than the value of the services that it provides to us. We are often not required to pay the costs of the air pollution that we produce when we drive and heat our homes; we pay the costs associated with supplying the energy but not those of the by-products of its use, and so we pay less than the full costs of using the energy.

The market price of natural resources does not include those resources' important roles in healthy ecological systems. This discrepancy between price and the full ecological values of resources in nature has contributed to many environmental changes. We live in places where lack of water would have prevented habitation in earlier centuries, places accessible only by car, places too cold for many people to live without modern heating technology. And both the environment and our values are different as a result.

A Changing National Personality

Leaving the places where we grew up, the companies that we work for, our families, and our work are more than just practical changes. These actions signal changes in values like loyalty; changes in identity, since people base their identities in part on where they live and where they work; and changes in responsibility and independence.

Loyalty means standing by someone. Leaving is the opposite, and we leave the places where we grow up more often than almost any other non-nomadic group of people ever has. We may make up for it with phone calls or e-mails, but those forms of quick communication do not represent the kind of fulfillment of binding commitment or loyalty that many people seek in their lives. We leave our employers more than any other group who has ever lived; we move out of our family home, on average, earlier than in most societies. This could be interpreted as a lack of allegiance.

It does not mean that people don't care about each other. It does not mean that they don't help each other. But it does mean that sometimes old friends are gone when people look for them, and that Americans have to readjust frequently to new people. It is part of what makes us American. It provides us with new opportunities for exciting careers, travel, new experiences. But loyalty to corporations, places, and others is diminished by separation.

We are more likely to stand by a place that we identify with. We are more likely to invest in a neighborhood that is our own. We might work harder for a corporation that represents a life-long commitment for us. Our backs are to the wall when we fight for our own places and jobs; but our backs are not against the wall if we can just as well move on to a new place. If we are less attached to places and companies, then this is a form of reduced loyalty. When we do not know the people who we do business with, they are more easily replaced than they would be if we had a long-term relationship with them built on trust or mutual interest. We are less loyal to them.

Under these frequent moves and rapid changes, it is difficult to make lasting decisions. A worker may have three, or five, or more careers during his or her lifetime. Is it wise to invest in years of education for any of those passing professions? A person may well live in more than a dozen homes. How much work should go into any one of those homes? If individuals have less commitment to the places where they are at the moment, or to the people who they live near at the moment, how hard will they try to become acquainted with their neighbors?

Personality springs from a person's beliefs and commitments. In part, it is about what the person will stand by and defend if the need arises. Do people who move often and who let their friends drift away have less personality? Are Americans losing responsibility as they find that they do not have to stand by their employers, their homes, or even the friends of their early years? If we focus on possessions rather than relationships, will this affect our ability or our willingness to compromise with others and bend to accommodate their needs as well as our own? If we focus on possessions, then will we tend to look at nature as a collection of objects for us to use rather than as a rich, integrated ecology that is most valuable when it remains intact?

Statistical rates of departure from houses and jobs and regions and membership organizations raise the possibility that the national personality is changing. Robert Putnam finds evidence of this change among generations of people:

> [Compare] the generation born in the early 1920s with the generation of their grandchildren born in the late 1960s. Controlling for educational disparities, members of the generation born in the 1920s belong to almost twice as many civic associations as those born in the late 1960s (roughly 1.9 memberships per capita, compared to roughly 1.1 memberships per capita). The grandpar-

ents. . . . vote at nearly double the rate of the most recent cohorts (roughly 75 percent compared with 40–45 percent), and they read newspapers almost three times as often (70–80 percent read a paper daily compared with 25–30 percent). And bear in mind that we have found no evidence that the youngest generation will come to match their grandparents' higher levels of civic engagement as they grow older.[59]

A country populated by this younger generation is a new world. Reduced newspaper readership means fewer common subjects to discuss. Reduced voting rates mean less interest in discussing politics or public affairs. Reduced membership in civic associations means less cohesiveness among people.

We have reduced other interactions as well, including our interactions with nature. We have moved into offices and condominiums, away from the cold or the heat, from most smells, from physical discomfort, and from other realities that our ancestors faced throughout history. People value what they are in touch with more than they value abstractions with which they have lost touch. Our departure from the natural environment makes it more acceptable for us to pollute it. Our experience of wood mostly as furniture, and of food mostly as pre-prepared, makes it easier for us to harvest virgin forests and let soil erode off of hillsides. Time spent downtown among concrete and honking horns, and in high-rises, changes people's view of their world. It changes the ways that we approach decisions about how to care for the environment.

In some ways, we have also moved ourselves away from other people. Office walls keep out other people as much as they keep out the cold. Spacious suburbs are about moving away from crowded cities, from urban problems, from traffic jams, from homeless people. Gated communities are not built to keep out the air or water—they are built to keep out other people. Sitting in a car means keeping away from other people more than we can keep away from them when we sit on a train, ride our bicycles, or travel on foot. Videocassettes offer entertainment in private, instead of in a theater of people crunching popcorn or sitting in front of us. Landlords live away from the buildings they have responsibility for, and so avoid suffering from problems in those buildings. Stockholders have little or no contact with the corporations that they own other than a strictly financial stake.

Our ability to move away from those things that we wish to avoid means a great deal for our public affairs. A population that has learned how to avoid toiling in the fields, and that has driven to the suburbs to get away from inner-city crime, is likely to believe that it can drive away from other problems as well. Its approach to homelessness will not be to reach the roots of the problem, but rather to sweep the homeless away into the next county, or downtown, or into a prison. Its approach to environmental problems will be to landscape its own lawns while allowing more distant ecology to decline. Moving away from problems is an approach to public policy. It is an approach that we are using often.

It is said sometimes that all politics is local. This cannot be the case in America. This is because there is no local. In Manhattan, the District of Columbia, San Francisco's Financial District, Phoenix, and myriad other locations, residents do not remain long enough to develop local politics. If an important election is held once every two years, then in Tucson, Arizona, the majority of all citizens have lived there for only two such elections. And the majority may be gone before two more such elections take place. Passion over local control of resources is meaningless to people who do not really have more than a temporary attachment to the place where they currently live.

These changes are part of our national personality. We are building a country that is built on ephemerality. Dependability is not the cornerstone of our values. Flexibility is. Devotion to public issues is not what drives our activities. We do not have to justify our actions to very many people, because we are not tied to those people. Moving away—in all its varieties—is a national behavior, and it is shaping our country. We are even moving away from the national personalities of the past.

New Connections

While Americans have moved away from many parts of their lives, they have also built new connections along the way. These new connections are defining much of what it is like to be American. They are also building many of our opportunities.

According to demographer Patricia Gober,

> No longer is Mexican food limited to the Southwest, jazz to New Orleans, country music to Nashville, and pizza to Italian neighborhoods. They are now aspects of a larger national cultural experience and, indeed, are being diffused worldwide. As more people gain the experience of living in different places, we see the erosion of regional distinctiveness and more uniformity in ways of speaking, eating, dressing, and doing business.[60]

These are cultural connections among people who have wandered away from their ancestors' ethnic divisions. Mobility diffuses cultural traits, spreading connections among people who previously did not share any aspects of their ways of life. It defines a new, national way of life, and of eating, that has connections into many cultures. The building of these new connections has long taken place in the United States. The movement of southern blacks to the North; the migration of Appalachian whites to Detroit, Cincinnati, and other cities, and of midwesterners to California and New Yorkers to Florida all established new connections.

We are even becoming connected to each other in completely new ways. Television signals connect us and spread common familiarities with fictional characters and celebrities. Many of these common connections are increasing, not decreasing. Tens of millions of people watched Jerry Seinfeld on television before his final show. They could talk about it at bus stops and diner counters. Telecommunications specialists are bringing cell phones to individuals who can now, if they can pay the bill, talk to others anywhere in the world. E-mail networks let people do the same while paying a smaller bill. International bureaucrats are working every day to link together more people through international trade and investment. These are all the building of new connections.

Water aquifers connect us as well, and few people would want to live without that connection. The roads are a connecting device among us, and one of the places where we can see other people who have little else in common with us. Even chain stores, in identical strip malls, connect us with a common experience. Radio signals, bookstores, music stores, airplane schedules are all ties. The electricity grid is a connection that did not exist before the twentieth century, and which brings common opportunities and warmth and information into the houses of people who otherwise have little in common. While we have moved away from many traditional connections, we have nevertheless created these new connections.

International organizations connect many of us together as well. The World Trade Organization, for example, works to encourage the free flow of objects and money between us and other countries, making a connection. The open trade routes that result update a connection that goes back to the Silk Route and before. Shared mutual funds are a connection. Each of these connections has its own effects, its own harms and benefits. Some bring new opportunities and health to us; others undercut our attempts to protect nature and provide secure jobs for each other.

For the most part, though, these kinds of contact do not give people much to base their relationships on. We watch the TV but do not participate in it. We see each other on the road, but rarely speak. We share the same water but do not join together very often to protect it or improve its quality. We shop in the same chain stores and listen to the same music, but these things do not define our characters or really make us part of a strong community.

These are connections between individuals but not bonds. They tend to be anonymous and nonparticipatory. They do not require us to engage each other actively. We are parts of international trade routes through our purchases and our money, but not in ways other than in deciding what to buy. We are connected through the electrical power networks; the electrons are a common denominator of modern human needs but convey nothing of human identity or aspirations. From a human point of view, these connections are devoid of real commonality.

The reduction of ties to the places where we grew up, substituted for by new connections made over the television or to products from other places, makes our world more anonymous. We are connected to people without knowing them. Meanwhile, the people who we know are often far away, and so less connected to us on a day-to-day basis.

Why does *People* magazine sell so many copies? In part, because it seems to redress this lack of connection. *People* magazine gives us information about people we know in common—celebrities. We don't know the people around us, but we do have a set of common acquaintances.

Unfortunately, we don't actually know these celebrities. It is an attempt to build an illusion of human connections. The ways to build real connections would be through common memberships in organizations and by conserving our friendships through our whole lives. We pay $3.95 at the grocery store checkout line for an imitation.

The fulfillment of this need does not stop with *People* magazine. No stronger a literary bastion than the *New Yorker* fills some of the same functions. It does not choose the actors or actresses of action films or thrillers very often, but it does choose a set of celebrities to write about. It chooses those on the fringe of politics or business or the fine arts. It often chooses the relatives of historical celebrities, politicians, artists. These magazines are trying to build a national gathering place, albeit for people who have never met each other in person. They are responding to our loss of connection to the people around us. People who live alone can turn to this community if they become lonely. Parents can count on these magazines after their kids reach 19.5 years of age and move out of the house. As we sit on the airplane on our way to start a new job in a new city, we can pull out our favorite magazine and join its community of famous or semifamous figures. We can meet new people and forge new relationships on the pages of the *New Yorker* and many other magazines.

We can turn to our televisions sets for the same purpose. The president of Fox Studios defends television against the charge that it is destroying community. TV, he claims, is stepping in to fill a public space that was already vacant. Americans' frenetic motion to new cities, new social groups, new jobs has evacuated our former gathering places. Our emotional and practical needs for those places and the bonds they create is being filled by TV, he asserts:

> . . . for decades, network television has been more successful than any one or any thing in fostering a sense of unity and community within our nation. . . . Over the years, . . . classics like *The Ed Sullivan Show, I Love Lucy, Bonanza, All in the Family, Dallas, The Cosby Show, Roseanne* and *ER,* among others, would provide this country with that same sense of a shared experience, something we could all identify with, and something we could all talk about, whether we were at the dinner

table, around the water cooler, or on the phone with our
friends who lived three thousand miles away.[61]

This TV executive may be right. If he is, then our country has reached
a surprising state. Amidst all of our success at designing new technologies,
at expanding our economy, at educating our people, we find ourselves
turning to television and magazines to fill a void as basic as our need to
belong to a community. Before we had acquired such a large GDP and
such impressive technology, we were able to meet some of our human
needs by just sitting on the front porch with the neighbors. The image of
rocking chairs and lemonade, and mosquitoes, on the porch in the
evening air remains with us still, even in big cities where few people can
sit outside. It endures because it is an important and basic image.

At a time when we move away frequently and easily from our child-
hoods, the neighbors who we grew up near, old jobs, and old friends, we
face the challenge of restoring some of what we have given up. Television
and magazines are not as strong a solution as the rebuilding of our civic
institutions, our clubs and associations. It may be easy to replace our
membership in neighborhood activities, like sitting on the porch or joining a school parent–teacher association. After all, most of us are surrounded by people. This restoration may soon become as important a goal
as designing new technologies or raising our GDP.

Notes

1. Patricia Gober, *Americans on the Move,* Population Reference Bureau, Population Bulletin, November, 1993; Patricia Gober, "Americans Losing Their
   Get-Up-And-Go?" Population Reference Bureau News Release, January 13,
   1994.
2. Ibid.
3. U.S. Bureau of the Census, "Geographical Mobility of People One Year Old
   and Older, By Sex, between March 1996 and March 1997," *Current Population Survey,* March 1997, http://www.bls.census.gov/cps/pub/1997/mobil-
   ity.htm.
4. Robert D. Kaplan, "Travels into America's Future," *Atlantic Monthly,* July
   1998.
5. U.S. Bureau of the Census, op. cit.

6. Gober, *Americans on the Move*; Gober, "Americans Losing."

7. Kaplan, op. cit.

(8.) Gober, *Americans on the Move*; Gober, "Americans Losing."

9. Ibid.

10. Ibid.

11. Ibid.

12. Ibid.

13. Frances Goldscheider and Calvin Goldscheider, Population Reference Bureau, *Leaving and Returning Home in Twentieth Century America*, Population Bulletin, March 1994.

14. Ibid.

15. Ibid.

16. Ibid.

17. Thomas North Gilmore, *Making a Leadership Change* (San Francisco: Jossey-Bass, 1989).

18. Ibid.

19. "Median Job Tenure Now Three Years," *San Francisco Examiner,* September 13, 1998. This is a survey of 11,771 adults.

20. Michael Lewis, "The Joy of Quitting: Ditching a Job Is Today's Secret of Success—Not Just in Silicon Valley, but in Washington Too," *New York Times Magazine,* September 5, 1999.

21. *San Francisco Examiner,* op. cit.

22. Maryann Cusimano, ed., *Beyond Sovereignty* (Boulder: St. Martin's Press, 1999).

23. Steven J. Davis, John C. Haltiwanger, and Scott Schuh, *Job Creation and Job Destruction* (Cambridge, Mass.: MIT Press, 1996).

24. Ibid.

25. Walter Russell Mead, "Planes, Trains, and Automobiles: The End of the Postmodern Moment," *World Policy Journal,* Winter 1996/97.

26. Ibid.

27. Irene M. Kunii, "Will Technology Leave Japan Behind?" Business Week, August 31, 1998.

28. Mark Kingwell, "Fast Forward: Our High Speed Chase to Nowhere," *Harper's Magazine,* May 1998.

29. Saul Hansell, "Is This the Factory of the Future: On the Internet Clock, Middlemen Are Turning Into Manufacturers," *New York Times,* July 26, 1998; Pete Engardio, "Souping Up the Supply Chain," *Business Week,* August 31, 1998.

30. Ibid.

31. Louis Uchitelle, "Global Tug, National Tether: As Companies Look Overseas, Governments Hold the Strings," *New York Times*, April 30, 1998.

32. Kingwell, op. cit.

33. Gober, *Americans on the Move.*

34. Louis Uchitelle, "Some Economic Interplay Comes Nearly Full Circle," *New York Times*, April 30, 1998.

35. Ibid.

36. Durwood Zealke, David Hunter, and Paul Orbuch, *Trade and the Environment* (Washington, D.C.: Island Press, 1993).

37. Robert D. Putnam, "Bowling Alone: America's Declining Social Capital," *Journal of Democracy*, January, 1995; Robert D. Putnam, "The Prosperous Community: Social Capital and Public Life," *American Prospect*, Spring 1993.

38. Ibid.; For additional data on related trends and surveys see also Theodore Caplow, Howard M. Bahr, John Modell, and Bruce A. Chadwick, *Recent Social Trends in the United States 1960–1990* (Montreal and Kingston: McGill-Queen's University Press, 1991).

39. Putnam, "Bowling Alone."

40. Ibid.

41. Nicholas Lemann, "Kicking in Groups," *Atlantic Monthly*, April 1996.

42. Ibid.

43. Putnam, "Bowling Alone."

44. John Cassidy, "Who Killed the Middle Class?" *New Yorker*, October 16, 1995.

45. Ibid.

46. Ibid.

47. Ibid.

48. David Cay Johnston, "Gap between Rich and Poor Found Substantially Wider," *New York Times*, September 5, 1999; Kristin A. Hansen, *American Demographics*, January 1998.

49. Mead, op. cit.

50. Clay McShane, *Down the Asphalt Path: The Automobile and the American City* (New York: Columbia University Press, 1994).

51. Ibid.

52. Mead, op. cit.

53. Ibid.

54. James Howard Kunstler, "Zoned for Destruction," *New York Times*, August 9, 1993.

55. Kaplan, op. cit., citing Marc Reisner, *Cadillac Desert.*

56. Unpublished manuscript from Trout Unlimited, Washington, D.C., 1998, and Water Watch, Portland, Oregon, 1998.

57. Kaplan, op. cit.

58. Wendell Berry, "Conservation and the Local Economy," in *Sex, Economy, Freedom, and Community* (New York: Pantheon Books, 1992).

59. Robert D. Putnam, "The Strange Disappearance of Civic America," *American Prospect*, Winter 1996.

60. Gober, *Americans on the Move.*

61. Sandy Grushow, president of Fox Studios, commencement address to the 1998 UCLA communications studies graduates, Los Angeles, May 1998.

The house has two floors; one above, where the corn cobs are sorted (we call it the tanpanco) and another below where we all live. But at the times of the year when there's no maize, many of us go up and sleep in the tanpanco. When the cobs are stored there, we have to sleep on the ground floor.

We don't usually have beds with mattresses or anything like that. We just have our own few clothes and we're used to being cold, because the roof doesn't give much protection. The wind comes in as if we were out on the mountain. . . .

Of course, when we sleep, we sleep like logs, we're so tired. We often get home so tired we don't want to eat anything, or do anything. We just want to sleep.

[When] we all go together to [live and work on a farm] . . . it's even worse for sleeping than at home, because we're with people we don't even know and there are hundreds of people and animals sleeping together. It is really difficult there. We're piled up in one place, almost on top of one another. . . .

In our case, all the brothers and sisters in our family slept together in one row. My older brother, who'd been married for some time, slept with his compañera, but the ones who weren't married (my other two older brothers, my sister, myself, and my three younger brothers who were alive then) we all slept together in a row. We put all the women's cortes together and used them for blankets. My parents slept in another corner quite near us. We each had a mat to sleep on and a little cover over us. We slept in the same clothes we worked in.

—Rigoberta Menchu, *I, Rigoberta Menchu:
An Indian Woman in Guatemala,* 1983

CHAPTER THREE

Living Alone

*Americans Have Acquired Something That
Few Have Ever Had Before*

Our tendency to move frequently to new places, to new jobs, to different environments, and away from some of the people we grew up among has many results. One of those results, however, makes the United States unique in the world, and has affected our lifestyles, civic culture, and country dramatically. It is that we have perfected a new kind of household. We now have a large group of people who live outside of any family structure, and without anyone else in their homes.

These people have acquired something on a scale unprecedented in history—living spaces with privacy. Large parts of the world consist of shantytowns with extended families living in spaces so small that even just one American citizen might demand more. The United Nations reports that 1 billion people have inadequate shelter. Even in wealthy parts of Europe and Japan, few people have the space or privacy afforded by an apartment of their own, or a house of their own. But in the United States, the trend is for people not only to live in large, private spaces, but for many people to live all by themselves.[1]

There was once a time when almost no one lived alone. In the United States, back in 1790, only about one-half of 1 percent of all citizens lived by themselves.[2] Some of them were explorers, who struck out alone. Others were loners who did not appreciate company. Some were widows and widowers. But everyone else lived in a world of kids and grandparents, of

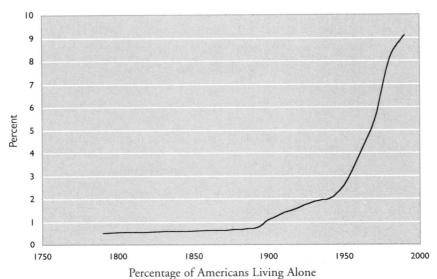

Percentage of Americans Living Alone

Source: Calculations by Craig Rixford and Cliff Cobb based on U.S. Census

snoring overheard in the night and cooking shared among the extended family, neighbors, and freeloaders.

Even 100 years later, the proportion of Americans who lived alone had scarcely risen. In 1890, the people who had no roommates or family members living with them still numbered scarcely more than one-half of a percent.[3] In total, they numbered 457,000 people. Around that time, though, the proportion began to rise. At about the turn of the twentieth century, it reached 1 percent for the first time. Even so, one in a hundred people living alone still constituted a fringe group.

By 1930 the proportion was getting close to 2 percent. That was a doubling in thirty years. It would not be the last. By 1960, the number had more than doubled again, and reached 4 percent of Americans living alone. By the start of the 1970s, more than 5 percent had their own place. One in twenty people begins to constitute a social group. Living alone becomes an accepted choice of lifestyle.[4]

In 1980, fully 8 percent of Americans had joined this group. And by 1990, the time of the last census, it was 9 percent. They numbered more than 22.5 million people. Starting with fewer than half a million individuals 100 years earlier, and with only 21,000 individuals in 1790, more Americans lived by themselves in 1990 than the entire population of Scandinavia. No new figures will be available until the 2000 census is compiled and reported. If the rise has continued at the same pace, though, we will have reached 10 percent. In any case, the United States has by far the largest proportion of people to live alone in any society at any time in history.[5]

Roughly twice as many people live alone now as did in the early 1970s, and about five times as many people live alone now as did at the beginning of the 1930s. They can put whatever they want into their refrigerators. They can come home at any time of the day or night. It is freedom in the most direct and personal sense.

This change has involved so many people that it deserves a place among the earthshaking changes of our time. But it has not gotten much attention. It is hidden in the history of the United States, though it may well have had a greater effect than politics in shaping our national character. It has left our evenings quieter, and restructured our personal relationships (and been restructured by our relationships), replacing large family dinners with dates and replacing two-person dinners with dinners alone. In some cultures, children and adults are punished by having to spend time by themselves. In America, people are spreading themselves out, and moving in alone, by choice.[6]

Like the quickening pace of our lives, these changes snuck up on us. While we were distracted by other stories in the newspapers and on TV, many people moved gradually to their own apartments and houses. Others saw their spouses pass away and found themselves living alone. Many people had many reasons, but taken together their situations have major implications, not only for individuals, but for public affairs, the environment, and the future. In some ways these changes are part and parcel of prosperity, the economic wealth that can buy larger homes, summer homes, and apartments or houses for one individual. In other ways, they mirror patterns of divorce, late marriage, freedom not to marry at all. They represent longer life spans for widows and widowers who live alone after the passing of their spouses. They even represent the changing demographics of the baby boom and baby bust.

Most of all, though, they represent American preferences and the ability to fulfill them. American kids can go off to college or move out of the home at eighteen, even when expensive tuition is required, a freedom wished for by their counterparts in Europe, Japan, and throughout the world who mostly continue to live with their families. American grandparents live in retirement homes, where nurses, instead of children and grandchildren, care for their needs. This lowers household size. Cousins no longer live with their extended family as much as they once did. Freeloaders are not welcome for long, even though there have been times when freeloading was a part of the national culture.

This is a world far removed from the America of the past, or from European, African, and Asian mores. Wealthy Japan has only about one-half as much residential floor space per person as in the United States.[7] But many Americans have made privacy and personal freedom two of their most desired commodities. They value quiet and order over kids and cousins, and have been among the first people ever able to translate those values into reality.[8]

These trends in living patterns may explain in part America's high rates of television viewing, as people turn to the TV for company. Larger numbers of one-person households increase the need for telephone lines, because people can no longer share a line (see figure graph of telephone lines in Chapter 1). These changes can raise telephone bills, as people hold their conversations long-distance. They may be part of the reason why electronic mail is suddenly so popular (see graph of Internet host computers in Chapter 1). They help to fund the airlines, as people travel to visit the family members who they have moved away from (see graph of world air travel in Chapter 1). The trend of shrinking household size changes dinner table conversations, making them more adult in homes with no children, or more introductory among people just getting to know each other. These changes build self-sufficiency in some. They cause loneliness in others.

Even if the trend toward solitude involved no more changes than those just mentioned, it would still be worth noting alongside the most striking political and economic developments in the United States today. But it causes other changes as well. Living alone is not only a social trend, it is the root of some of the physical changes taking place in the United States. On the weekend, many people don't stay home with family—they get in the car and drive to the place where their family lives. This causes traffic and gas consumption. Roads are built to accommodate. Whole industries are created or grow to accommodate—the greeting card industry, overnight mail, the building of retirement homes, the building of college dormitories and cafeterias to house and feed the people who have moved away from their families for the first time, the building of much of America. This is one of today's trends that tie social and environmental issues together.

This choice goes hand in hand with a larger consumption of natural resources to build, heat, light, and maintain homes than has been spent in any other country except Canada. The United States is the world's largest producer of carbon dioxide and consumer of fossil fuels, for example, in part because we heat larger living spaces and drive greater distances between them. Building more and larger homes requires us to pave more land, to quarry more stone, and to add all of the plumbing, electrical wiring, and other components of modern houses and apartments. Once they are built, these homes must also be heated, cleaned, and maintained. As a result, this is a choice to use more oil and gas, to pour concrete over new areas, and to affect the environment more broadly. Even though consumption of coal, the dirtiest of the fossil fuels, is falling in most parts of the world, it is rising in the United States, by 1.7 percent in 1998 for example.[9] (See figures in Chapter 1 for the paving of roads and consumption of petroleum and for U.S. coal consumption.)

Solitary living is the most extreme form of a set of changes that have expanded the average amount of living space per person and created many households of two or three people, spread them out into subdivisions, and remade the household experience. This is a social experiment that has never been tried before in history. The jury is still out. Much of what this undertaking will mean for its participants, for their distant families, and for the environment remains to be seen.

When the founders of the United States spoke of freedom, they did not mean freedom from family obligations. The parts of our Constitution that guarantee liberty are not referring to the liberty to sleep late in the morning, or to make noise late at night. The "pursuit of happiness" conceived by early political philosophers did not revolve around the ability to decorate a home according to one's own tastes. But over the past few decades these are among the freedoms that Americans have pursued the most energetically.

Saying Goodbye to the Relatives

The size of an average U.S. household has been shrinking for 200 years. In 1790, it was 5.8 people. In 1890, it was 4.9 people. That drop of almost one family member over the course of a century was a rapid change by the standards of previous centuries, when people had fewer opportunities to move to new homes, jobs, or locations. But the changes were just beginning. By 1960, average household size was just 3.3 people. In 1993, it was 2.6 people.[10]

When we speak of the household today, we mean something different from the homes described in the documents of our history books. We mean something that is less than half as full of people. When we talk about going home at night, we are referring to a different experience from the one that characterized American life during much of this country's existence.

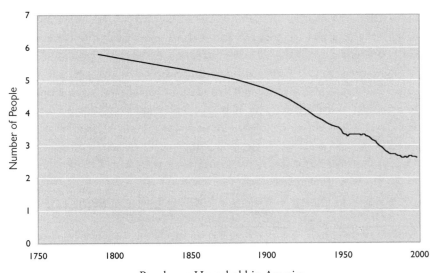

People per Household in America
Source: U.S. Bureau of the Census

Is it possible that this is the primary change in the American experience over time? With fewer family obligations, people can stay in school longer and so pursue new career opportunities. They can concentrate on their homework in relative quiet. They have the time to pursue hobbies and pastimes and receive midlife training for second or third careers. They can work longer hours: between 1977 and 1997, the average work week among salaried Americans lengthened from forty-three to forty-seven hours, according to *U.S. News & World Report*.[11] The Families and Work Institute reports that the number of workers putting in 50 or more hours a week jumped from 24 percent in 1977 to 37 percent in 1997.[12] These changes not only affect our homes, they also give us opportunities to learn new careers, to socialize with new people, to travel without responsibility for others, and to have the freedom to try myriad new things. This shrinking household population is the transformation of the experience of the extended family into the experience of modernity.

Americans have done voluntarily what the governments of some other countries have tried to accomplish through politics, or even through force. The Chinese government has tried—hard—to shrink the size of Chinese families and households. It limits each family to having just one child, and enforces that policy with the threat of imprisonment and other punishments. (Ethnic minority families can have two children.) But Americans skipped that political dragon. We went beyond such politics. For us, smaller families were part of the way that we responded to the economic and social issues of our time, and the way that we wanted to shape our own personal lives. No political debate took place—it was a personal choice that, inadvertently, projected itself into public change as well. Many countries in the developing world now face a need to slow their dynamic rates of population growth. It is one of their most important challenges; but in America it happened without intention.

Our way of life has changed as well. Scarcely a decade ago, Americans looked with amazement at the long Japanese workweek. But now we have passed the length of that workweek ourselves, with the International Labor Organization reporting that we work more hours per week than any country in the industrial world.[13] This is not due entirely to changing household sizes. But long working hours have affected such topics as divorce and the urgency felt by young Americans to begin their careers;

and smaller households have made more people able to put in long hours at work. Those Americans who do live with their families still say goodbye to their relatives in the morning when they go to work, and on average they don't say hello again until a later hour in the evening than they did in decades recently past. (It is worth noting that in the more distant past, around the beginning of the twentieth century for example, many Americans worked ten-hour days and six-day weeks in factories, and even longer on farms, and then went home to difficult housework.)

In the recent trend of longer workweeks, we have moved farther away from the characteristics of Europe. The average American works the equivalent of eight weeks a year longer than the average Western European today. France has a mandatory thirty-five-hour workweek promoted by the government as a way to cut unemployment. In Norway, Sweden, and other parts of Europe, ordinary workers get four to six weeks of vacation plus holidays.[14] They can spend more nonworking time with their relatives than we can in the United States.

Americans have accomplished this transformation quickly. During the first hundred years, our average rate of decline of household size was about 0.16 percent per year. But from 1890 to 1960, it proceeded at 0.56 percent per year. From 1960 to 1990, the household shrank at an average rate of 0.8 percent per year. In the 1970s the rate of decline peaked at 1.3 percent per year.[15]

In the 1960s and early 1970s, the country's public attention was focused on social criticism around the war in Vietnam and the rebellions of young people. But the largest social change taking place at that time may have gone almost unsung, as households, and household responsibilities along with them, shrank at 1.3 percent every year while Bob Dylan played his guitar, and at nearly that rate while Janis Joplin challenged cultural mores. It was not the rebels or the hippies that made this change, but rather the mainstream bankers, factory workers, and college students who moved out of their parents' homes and exercised preferences for smaller households.[16] It is ironic that those people who poured their hearts into calls for social change may have accomplished less change than the people who inadvertently contributed to the decline in household size. If the counterculture fed this trend, then it was not through slogans or songs but rather in the act of rejecting the nuclear family in the protesters' own lifestyles.

Family and household sizes fell for many reasons. Back in the years from 1790 to 1890, small households were rare. No one else in a neighborhood had much privacy or space, so it did not occur to people to wish it for themselves. Children and cousins were indispensable labor for doing chores around the farm and around the house. Most likely, few people even thought about having smaller families. If someone had told them that the future would hold such small families, people back then would have wondered who would do the work, and what all the excess space would be used for.

Several changes took place, however, to make smaller households a possibility, and once the momentum gathered, clans began to shrink. The country's long-standing move away from family farming made smaller households more plausible because fewer family members were needed to do farm chores. Manufacturing and specialization added to productivity and fed the market economy in which people could pay for skills such as repair work, clothes making, cooking, plumbing, and other activities formerly done by family members. The move toward manufacturing that delivered ready-to-consume products, from clothes to meals, made some household duties, like spinning and weaving, almost obsolete. Previously, husbands, wives, kids, and cousins played key roles in household maintenance and on farms. When the changing economy eliminated some of those roles, it freed more people to live in smaller households.

Modern birth control also became available, which affected household size powerfully. And rising incomes, especially in the Roaring Twenties and the years after World War II, made living alone more affordable. Eventually, by the 1970s, the time when households shrunk the most quickly, social freedoms were proliferating and economic growth was making those freedoms possible. People both wanted to, and could afford to, live in new ways.

Shrinking household size is also about marriage. In 1940, the ratio of married couples to people living alone was 10 to 1. By 1992, that ratio had fallen to just over 2 to 1. This is a reflection of the fact that the marriage rate in the United States has fallen considerably over the years, from 93 marriages per thousand single women in 1970 to 58 today. For single men, the comparable figures are 80 and 48. For both men and women the average age of first marriage moved back by 3 years during the same time period, from 20.6 years to 23.7 for women and from 22.5 years to 25.5 for men. Some of these people share living quarters with family or house-mates before they marry. But many of them live alone. This is part of the changing American experience.[17] With fewer marriages, birth rates also fall, which reduces household size still farther. And as other individuals and families also have fewer children, there are fewer live-in cousins as well, and even fewer freeloaders from other families.[18]

Another factor that explains how so many Americans have come to live by themselves is divorce. The country's divorce rate doubled between 1965 and 1975. Today's divorce rate has actually fallen slightly from that of 1975, but still remains twice as high as it was in the 1950s. It now stands at 4.7 divorces per 1,000 Americans, exactly half of the 9.4 people who got married each year per 1,000 Americans lately. Among people who got divorced in the late 1960s, 60 percent remarried within three years; but by the early 1980s, only 36 percent did. Such a sudden and dynamic rise in divorce has reshaped many people's lifestyle, relationships, and living patterns.[19]

In 1910, only about 12 percent of the elderly lived alone. Today, 40 percent of those over 65 live alone, and many are widows or widowers. In 1910, most of the widowed elderly lived with one of their children, but in 1990 most lived alone. The rise in living alone among the elderly occurred particularly between 1950 and 1980. Greater independence among the elderly, and retirement homes for them, also played a large role in chang-ing the places where they live. Many were both able and willing to spend more time by themselves, and others had the decision made for them by trends in mortality rates among their spouses.[20]

Since, statistically, women generally marry older men, and then live an average of seven years longer, particularly large numbers of elderly women live alone. The number of women living alone doubled between 1970 and 1998, from 7.3 million to 15.3 million. Half of the women living alone were elderly, and among elderly Americans who live alone, almost 80 per-cent are women. Put another way, 41 percent of all elderly women lived by themselves. Nearly half of elderly women are widowed, compared with just 14 percent of elderly men.[21] Their experience is a growing part of the American experience, and one that deserves more coverage in the mainstream media.

The post–World War II baby boom and the baby bust that occurred before and during the war affect how many elderly people live alone. Elderly people whose children were born during the baby bust of the depression years are living alone in large numbers because they have fewer children with whom they might live. The parents of boomers, meanwhile, are less likely to live alone than the previous generation because their larger numbers of children provide more possibilities for them to live with one of those children. Demographers predict that there will be an upswing in elderly people living alone between the years 2005 and 2010, when the parents of boomers have moved into their 80s or died.[22]

Many other changes taking place in America affect these trends of living alone, late marriage, and smaller household size. Schooling, for example, plays a significant role in determining whether someone will live alone or not. Average age at first marriage increases with levels of formal education—people with bachelor's degrees marry later, on average, than those with only high school degrees; those with graduate degrees marry later still; and those with Ph.D. degrees marry the latest, on average.

Since many of the researchers who study these subjects have advanced degrees themselves, they can joke that these trends are caused by the demanding nature of graduate school, or by the personalities of those who choose to write dissertations. Many people believe that the real cause of these trends is that people with advanced education replace some of the rewards that they could have gotten from having children and families with different rewards that they derive from doing work that they care about, traveling to conferences, and devoting themselves to their studies. Education gives people more options and opportunities—choices that they can follow that are different from the traditional path of raising children and working for the money needed to support children and spouses. Statistically, those who attend at least some college are more than twice as likely to live alone as those who have not attended college.[23] This makes education part of the march toward smaller households.

The relationship between education and living alone is strong enough that the national goal of increasing education is a de facto national goal of smaller household sizes. The U.S. Department of Education does not consider family size when it does its work of promoting literacy or knowledge; nor does Congress consider family size when it passes legislation dealing with education. But those activities have the effect of shrinking family and household sizes anyway. Indeed, they may be the most effective ways of all of shrinking families and households.

In other countries where it is a national goal to reduce population growth, one of the express priorities for accomplishing smaller family sizes is to increase the number of years that children spend in school, especially girls, who otherwise have lower literacy rates. At the 1994 United Nations Conference on Population in Cairo, lengthening children's education was chosen as a central part of the plan of action to reduce population growth.[24]

Perhaps U.S. high school students should receive a counseling session before applying to college, where a therapist will tell them that college will help them find great careers, earn more money, become well-rounded people, and meet many new friends—but that it is also likely to delay their own marriages and, statistically, might lead them to live alone far longer than people who don't attend college, perhaps even for their entire lives.

The trend of living alone has affected some parts of the country more than others. People who live alone tend toward certain locations. According to researcher Paul Glick, "places in the United States where an unusually large proportion of households consist of one person living alone include: parts of the Midwest farm belt with an aging population; rapidly growing areas with high immigration rates, like southern California, parts of Florida, and parts of Texas; areas where ample housing is available; college communities; cities making use of an above average proportion of workers in professional occupations; and localities where most of the adults feel relatively free of traditional family values." These places may be a foreshadowing of what is to spread into new areas in the future.[25]

Many American social preferences today promote small households. If schooling more than doubles rates of living alone then schooling can be considered an activity that shrinks households. Professional occupations promote small households by making money available for individual apartments and by making time scarce for sharing with others. College communities do as well, with their abundant rental units for individuals and plenty of services provided by moving companies that pass out brochures on the street. Rapid construction of new homes also contributes to the trend, with the additional space and housing units that it makes available. Immigration does as well, since some immigrants are not able to bring their families with them.

All of these are issues of living patterns, not just schooling, affluence, immigration, or building. In our politics and newspapers, we should acknowledge them as such. Yet privacy and isolation are not issues that have had a place either in political debates or in discussions of education levels, housing construction, or immigration. Instead, privacy and isolation are hidden parts of our politics. The descendents of the people who wrote the documents that gave us fundamental political freedoms way back in the 1700s have since turned the country into a place of freedom from family chores, of income freed for expenditures that do not have to do with children, of leisure time away from the relatives.

The Infrastructure of Privacy

Each person living alone makes up a whole household. That means that by 1990 almost one-quarter of the country's households were made up of individuals living by themselves. The nearly 10 percent of us who live alone represented nearly 25 percent of homes. That is about triple the proportion of one-person homes in 1950. It is almost five times the level of 1900. It makes people living outside of families into a "demographic" of their own—a group whose homes are targeted by advertisers and political candidates and served by government, and whose tastes are represented in the country's shops and dining establishments. The products and services that are created for these individuals constitute an entire infrastructure of places to live, foods to eat, and devices to save time that make it easier and more possible for Americans to live alone or in smaller households.

On average, people who live alone have more money for themselves than people who live with others. Average income per household member for those who live alone is $20,600. For married couples, average income per person is $14,600. For those who are not married but who live with other people, such as single parents, people with roommates, or unmarried couples, average income per person is $10,100. It's true that the

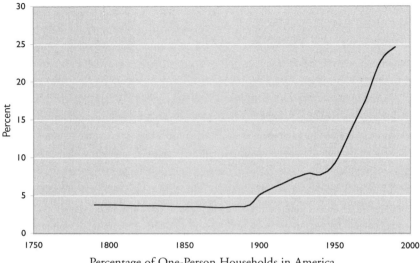

Percentage of One-Person Households in America
Source: U.S. Bureau of the Census

household of a married couple earns more money than a one-person household does, and shared housing situations with multiple people often do as well. But that income is shared among more people.

Those individuals who live by themselves have significantly more money for themselves. If one dollar equals one vote in the marketplace of items for sale, then those of us who live alone get more than our share of votes toward which products are produced and which services are offered to America. The tastes of those individuals play a powerful role in the worlds of retail, marketing, politics, and government.[26]

Having this group of people in the marketplace alters the national shopping spree. They influence the industries for housing construction, furnishings, entertainment, food, and travel. More of their products than ever before are created to appear attractive to people who have no family to keep them at home, and who seek opportunities and entertainment outside the home. No matter how well the advertising industry targets this "demographic," the tastes embedded in these products reach all of us over the airwaves of TV and radio, in print ads, and on billboards. Among other people, this may reduce contentedness with staying home with the family, since more products advertised and displayed are pitched to people who have no family.

The larger purchasing power per person among those who live alone—$6,000 more per year than members of married couples—makes them an appealing target for manufacturers and advertisers. And each person living alone needs his or her own furniture, home decorations, TV set, stereo, and other items that would otherwise be shared among the multiple members of a household.

Our restaurants, our gas stations, our grocery stores all respond to the tastes of small households and individuals. Take-out sandwiches, pizzas, and all sorts of fast food make people able to remain on the go, to eat affordably without going home, to work and keep house at the same time. The size of the U.S. pizza industry is $21.8 billion a year, as it serves both single people and families.[27] Other time-saving devices include dry cleaners and laundry services, maids, frozen foods, supermarket salad bars, all of which are attractive to individuals who have no one at home to help them with cleaning or cooking. They include fast-food restaurants, sidewalk coffee booths, and pay-at-the-pump gasoline, where credit cards or cash can be inserted directly into the gas pump, saving a trip inside to pay. These devices help people live in a way that does not require the shared support of a family. This is not to say that families do not use these devices—families use all of them. But these conveniences do serve to free some individuals from their familial past. Without the living alone demographic to purchase these services and products, the industries that produce them would be smaller.

Marketers respond to the shopping preferences of single people, and the results change the contents of our supermarket shelves and the menus of our restaurants. Single women spend more than single men on sugar, sweets, and shoes. Single men spend more on alcohol, new cars, and eating out, according to Patricia Braus of *American Demographics* magazine. In fact, single men spend much more on eating out than single women do. Bachelors spend $1,600 a year on restaurant meals, almost as much as the amount spent by the average 2.6-person household. That amount is twice what is spent on food away from home by single women. Single women spend less than $800 a year on meals out. The main reason for this difference is that a large portion of single women are elderly, and elderly people

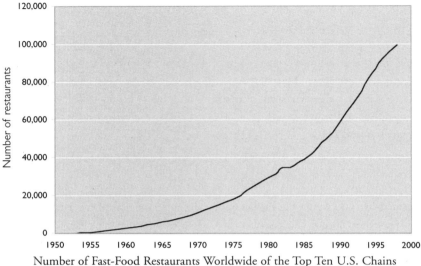

Number of Fast-Food Restaurants Worldwide of the Top Ten U.S. Chains
Source: Gary Gardner, Worldwatch Institute

are less likely than younger adults to eat out. Twenty years ago—and fifty years ago—when our country had many fewer single people, the industries that produce sugar and shoes, and alcohol and restaurant meals, were all quite different.[28]

The consumption choices of people who live alone even change our schedules, and the times of day when we can shop. It is possible that the flexible schedules of independent people, who may work in the day and shop in the evening, who do not get up to cook breakfast for the kids and do not have to have dinner on the table at 6:00, are part of the reason why the operating hours of many stores have lengthened over time. They may be the reason why shops are open longer in the United States than in any other country.

Many of the members of our fleet of freedom-generating devices are distinctly American. Pizza may have come from Italy, but pizza *delivery* came from the United States—and while it has arrived on a small scale in a few other countries, it only really has a niche in the country of solo living. (Many American pizza chains have recently added "personal pizzas" to their menus, smaller pies sized for one person alone.) Pizza delivery allows people to work and still have hot food at their doorstep. It supports those who have not learned to cook and those whose arthritis or other illnesses discourage them from cooking. Many of those people are among the ones who live by themselves, as some are young and have not yet learned to cook for themselves and others are elderly and have begun to have a harder time with the motions of cooking and shopping. Food of any kind can now be gotten by delivery. In many cities special delivery services exist from companies that do not make food but travel around town visiting a host of restaurants and carrying the food to people waiting at home.

Our living patterns affect which machines we own, and which machines our companies choose to produce. Some critical machines offer connections to loved ones and strangers. Without the telephone, fewer people would dare to be by themselves, for fear of loneliness, or the inability to reach other people if they need help. Likewise, cars make those who are alone able to get quickly to other people, and so cars make living alone more plausible. Now that computers and the Internet are coming into our homes, it is possible that some people who have wanted housemates or who have stayed with their families will feel a little bit more able to strike out on their own. Our growing independence in the home increases our appetite for some of these machines and swells the size of the markets for them.

The sales of these machines, and of services valuable to independent people, fuel economic growth. More of them means a larger economy, which is fueled by the needs of people living alone. And having the wealth that comes from that growth allows additional people to move off by themselves. Living alone is part of the cycle of economic growth in America. The privacy, isolation, and revised social interactions that it entails are all by-products of this economic growth. It means more construction work to increase the average living space per person. It means expensive dates in restaurants instead of meals at home. It is our personal relationship with free-market economics, one of the ways that we let economics into our private lives and one way that we use our private lives to shape our economy.

A major component of the infrastructure of privacy is space. It takes room for people to be able to go off by themselves. The median size of a new house built in the United States in 1949 was 1,100 square feet. Twenty-one years later, in 1970, it had risen to 1,385 square feet. Twenty-three years after that, in 1993, the median new American house covered fully 2,060 square feet. Once the data are available for 2000, we may see that the size of the median new American house doubled in the second half of the century.[29]

But this doubling of the size of our buildings does not accompany a rise in the size of families or households. On the contrary, it comes with a shrinkage in the people living in a typical household. In 1950, the average amount of residential space per American was 312 square feet. In 1993, it was 742 square feet. This is already far more than a doubling of space, and the trends of rising housing area and declining household population are both diverging still farther. Throughout the industrial world, average home size has risen since World War II, even as family size has shrunk. America leads this change.[30]

Our landscapes would look different without this rise in living space and such a large number of Americans living in smaller households. According to figures from the Sierra Club, between 1970 and 1990, Chicago's population grew by 4 percent—but home development coverage expanded by 46 percent and commercial areas grew by 75 percent. This expansion represents more living space per person, on average, and more shopping space as well. Cleveland lost 11 percent of its population—but still paved and built on 33 percent more of its ground. This is, once again, an expansion of living space per person, of people living alone, living with fewer relatives, or living with just as many people but in larger rooms. During the same period, the population of Los Angeles grew by 45 percent, but that was more than matched by 200 percent growth in land consumption.[31]

This expansion affects all of us, not only those who have more space themselves. Residential sprawl development costs more tax money to provide for public infrastructure such as schools, roads, sewers. In fact, it costs more additional tax money than it creates in revenues. The city of Fresno, for example, has doubled in population and size since 1980, mostly in sprawling suburbs, and this growth has resulted in $56 million in additional yearly revenues. But the cost of the services that the growth has required has been twice that much, and has risen by $123 million. This does not even include new capital costs like roads and sewers, which will cost more.[32]

As people spread out into larger living quarters, it changes our schools as well. In Minneapolis–St. Paul, between 1970 and 1990, 162 schools were closed in urban and central suburban areas, while 78 brand new schools were built in the outer suburbs. Similar trends took place with libraries, sewage and water facilities, and other infrastructure, both in these cities and in other cities and towns around the country.[33]

These expansions of living space affect our farms, our natural environment, and the views that we see from our cars and our windows. From 1970 to 1990, more than 30,000 square miles (19 million acres) of rural lands in the United States were developed. Every hour of every day, America loses 45.6 acres of its highest quality farmlands to subdivisions, shopping centers, strip malls, roadways—400,000 acres a year. Between 1982 and 1992, an area roughly the size of Vermont was developed. Our highest quality agricultural lands are often located close to metropolitan areas, because those are the places where people originally chose to settle because of the high quality of the land. It is precisely this land that is covered first by urban sprawl. These are not inevitable trends. They are the result of our choices for housing, our disinclination for sharing our space with others, and the economic subsidies and taxes that affect our decisions.[34] (For a long discussion of these subsidies and taxes, see Chapter 5.)

The increase in the space we consume, in the size of our buildings, and in the devices that we put inside these buildings ties our privacy to the environment around us. Cars and factories receive notorious attention as polluters. But the environmental harm caused by buildings has largely escaped scrutiny.[35] When tallied, the energy used to heat and maintain our buildings, plus the chemicals used in them every day, cause major ecological change. Likewise, the materials and energy used in their construction, and the land taken up by their size, are major contributors to environmental change.

Meanwhile, estimates of the number of homeless people in the United States are difficult to make. By conservative counts, though, the homeless number at least 300,000. These individuals contrast sharply with the fully 10 million Americans who have two or more homes. This divergence shows another component of the rise of living space and privacy in America: a rise in inequality of space. Because of its two extremes of lack of housing and abundant housing, America also has the dubious role of leading the world in abundance of privacy juxtaposed with unmet needs for the most basic facilities.[36]

The infrastructure of privacy means nothing to those who cannot afford any place to live at all. For them, the products marketed to people wealthy enough to have their own place only represent an additional distance between themselves and the tastes and goals of the rest of the country. This division accentuates the differences between those social groups too poor to live alone, or even to have basic privacy, and those wealthier people who take privacy for granted. The two groups live a different experience, one that is diverging farther as more people join the group who do not even share their housing at all. One group still lives the historical experience of shared sounds and spaces, in a crowded home, or even on the street. The other has left the past of scarce or shared housing entirely.

It is ironic that the infrastructure of privacy serves both of these groups of people—the homeless and the rich. People living on the street may eat a slice of pizza from the take-out window because it is inexpensive and anonymous and because they have no kitchen, while wealthy executives sometimes eat that same slice of pizza because it is fast and located near their next pressing meeting. The same streets where people carry their briefcases on the way to work are the places where the homeless put cardboard boxes to sleep in.

Changing Responsibility

Some people's obligations to their families have given way under this trend of solitary living and smaller households. Duty to children, parents, or grandparents is met by sending money in an envelope to the places around the country where they live. Afternoons out playing ball with a child give way to occasional telephone calls to the same child. Weekends fixing the house, building a tree house, or mending the fence can now be redirected to whatever activities the independent person chooses, or to longer working hours or more studying. Obligations that once kept young people home instead of out exploring the world have given way as children move out of the house at young ages and often live across the country from their parents. (See Chapter 2: Moving Away.)

In this sense, the trend of living alone or living with fewer people is a living pattern that makes people less responsible for others. The first sound they hear in the morning does not remind them to change the baby's diaper. They may not have to run to the pharmacy to pick up another person's medicine. The people who live apart may take on additional responsibilities at work or for a social club. But this partial freedom from family obligations has changed our experience. No law passed by Congress can match the fundamental nature of such change; no decision by an American court can affect people as deeply. The changing connections to the family are what make the American experience what it is today—both through the freedoms and opportunities of not being constrained by familial duties and through the losses of family life.

In a world where people have shed responsibility for family members and for extended family and freeloaders, they can turn their attention toward new freedoms and priorities. The freedom of the road comes more easily to those not tied down by family responsibilities, and so travel can increase. With fewer familial financial burdens, individuals can turn their values and pocketbooks toward material possessions. They seek privacy as a commodity, which they pay for through higher housing prices for larger apartments, or even through first-class airline tickets or box seats at a ballgame. They may try to make up for loneliness by buying impressive stereos or large televisions. They can throw large parties. They can squeeze in more movie watching, which they may need to assuage feelings of isolation. They have time for more exercise, which releases some of the same hormones otherwise released by intimate relationships between long-time spouses.

The American experience today is, statistically, one of more travel, more space, more material possessions, changing careers, and less family. This kind of living pattern involves significant anonymity. It means time spent away from the place where a person grew up or where they live, in new places where they may not know many other people. It offers room to be alone. Its versions of communication often mean time spent with e-mail or on the phone instead of meeting in person. Many people have the ability to abandon a tiring career and go to work in an entirely new office, or even to consult among frequently changing clients, to telecommute, or to work for themselves. It involves freedom from watchful eyes, from the need to uphold one's reputation, from family expectations.

Those who have anonymity may worry less about behaving in ways that would make them ashamed if they were among people they knew, like buying drugs, or even just spending less time helping others. They live in a world of different requirements and expectations. They have less responsibility to their neighbors if they do not know them. They have less responsibility to their community if they do not consider it to be their community.

This lack of responsibility may explain society's willingness to let its natural environment decline. What people learn at home they apply in the larger world. If they live independently, then their approach to dealing with nature may also be to assume that they are independent from it, even if they are not. People may have always believed that they should control nature and make it serve them. But if they learn that they have full dominion in their homes, then they may be more prone than ever to believe that they have full dominion over their natural surroundings as well.

A changing outlook on responsibility would explain many issues in society today. It would affect the public's views on homelessness. It might reduce political support for aid to foreign countries. It might make "donor fatigue" more common. It would lead to opposition to taxes for education from people who do not have children in school. It would lead to pressures for welfare reform or reduction from people who are not on welfare. Responsibility is not a single-issue characteristic. It crosses political boundaries, bridges political affiliations, and affects our response to many, or most, national choices. It affects all social issues.

For the environment, the pursuit of private living spaces and opportunities has brought change as major as the environmental change brought by the settlers who originally colonized America in search of freedom of religion, speech, and assembly. This change has come both directly and indirectly. In a direct sense, as living space per person has more than doubled since World War II, logging, mining, building, and energy consumption have all risen along with it. Rural areas became subdivisions near the places where American Indian settlements were once usurped by cities.

The deepest environmental changes, though, are those that come indirectly. They come from the changing of our other values, like responsibility, commitment, and connection, and all the results of the different choices that we make. Those changing responsibilities and choices touch almost everything we do. They affect what we expect from life; what careers we choose; whether we live our whole lives near the people we grew up among or whether we move away. They affect how much money we give to charity and, more important, how much of our time we share with our neighbors and our friends. There is little that they do not affect.

A person's personality is formed, in part, from learning to live with other people. It comes from hearing other people bang around in the kitchen, get angry at each other, sneak in late at night. It comes from learning to compromise in furniture arrangement, and from parents who refuse to compromise on values and principles. It comes from cooperation and conflict. People who grow up in smaller families and households may develop different personalities, different expectations for privacy and space, different ideas of shared responsibility, from those who grow up in larger households. Individuals who live alone in their twenties, or their seventies, may learn different behaviors and values from those who do not.

In countries where housing is scarce or where weak economies make sharing a necessity, as few people live alone today as did in the years 1790 and 1890 in the United States. This makes the United States an extraordinary country. It is not unusual just according to superficial differences, like use of acne cream or taste for fast food. Not even just according to average incomes or total size of its economy. It is remarkable for differences as fundamental as unprecedented levels of privacy, historically unique choices for how many people to live among, unprecedented inequality in access to space and amounts of sharing, and for all the machines and services associated with those differences.

Notes

1. "The Problem of Homelessness," in *Building for the Homeless* (New York: U.N. Department of Public Information, 1987).
2. Unpublished calculations by Cliff Cobb, Redefining Progress, San Francisco, Calif., 1997, based on data from the U.S. Bureau of the Census.
3. Ibid.

4. Ibid.

5. Ibid.

6. Barbara Holland, *One's Company: Reflections on Living Alone* (Pleasantville, N.Y.: The Akadine Press, 1992).

7. Lee Schiper and Stephen Meyers, *Energy Efficiency and Human Activity: Past Trends and Future Prospects* (Cambridge: Cambridge University Press, 1992) cited in David Malin Roodman and Nicholas Lenssen, *A Building Revolution: How Ecology and Health Concerns Are Transforming Construction,* Worldwatch Paper 124, Washington, D.C., March 1995.

8. Holland, op. cit.

9. Christopher Flavin, "Growth in Fossil Fuel Burning Slows," in *Vital Signs 1999: The Environmental Trends That Are Shaping Our Future* (New York: W.W. Norton and Worldwatch Institute, 1999).

10. Brad Edmondson, "Do You Want to Be Alone?" *American Demographics,* October 1994.

11. James Lardner, "World-Class Workaholics: Are Crazy Hours and Takeout Dinners the Elixir of America's Success?" *U.S. News & World Report,* December 20, 1999.

12. Ibid.

13. Ibid.

14. Ibid.

15. Unpublished calculations by Cliff Cobb, Redefining Progress, San Francisco, California, based on Edmondson, op. cit.

16. Ibid.

17. Cliff Cobb, unpublished research paper on living alone, Redefining Progress, San Francisco, California, 1997.

18. Paul Glick, "Living Alone during Middle Adulthood," *Sociological Perspectives,* Fall 1994.

19. Ibid.

20. Ellen A. Kramarow, "The Elderly Who Live Alone in the United States: Historical Perspectives on Household Change, *Demography,* August 1995; Ellen A. Kramarow, Research Report No. 93-287, Population Studies Center, University of Michigan; Merril Silverstein, "Stability and Change in Temporal Distance between the Elderly and Their Children," *Demography,* February 1995.

21. U.S. Bureau of the Census, from their Web site: <http://www.census.gov/Press-Release/www/1999/cb99-03.html>

22. Kenneth S. Y. Chew, "Urban Industry and Young Nonfamily Households," in *Housing Demography: Linking Demographic Structure and Housing Markets,*

edited by David Myers (Madison: University of Wisconsin Press, 1990); Linda Waite, Frances K. Goldscheider, and Christina C. Witsberger, "Nonfamily Living and the Erosion of Traditional Family Orientations among Young Adults," *American Sociological Review* 51 (1986).

23. Paul Glick, op. cit.

24. Lester R. Brown and Hal Kane, *Full House: Reassessing the Earth's Population Carrying Capacity* (New York: W.W. Norton, 1994).

25. Paul Glick, op. cit.

26. Cliff Cobb, "Living Alone: Basic Demographic Information," unpublished background paper, Redefining Progress, San Francisco and Sacramento, California, 1997.

27. Constance L. Hays, "What Companies Need to Know Is in the Pizza Dough," *New York Times,* July 26, 1998.

28. Patricia Braus, "Sex and the Single Spender," November 1993, Overland Park, Kansas: American demographics using Bureau of Labor Statistics' Consumer Expenditure Surveys (CE) for 1984–85 and 1990–91.

29. David Malin Roodman and Nicholas Lenssen, *A Building Revolution: How Ecology and Health Concerns Are Transforming Construction,* Worldwatch Paper 124, Washington, D.C., March 1995; The New Road Map Foundation, *All-Consuming Passion: Waking Up from the American Dream* (Seattle: New Road Map Foundation, 1993).

30. Roodman and Lenssen, op. cit.

31. Sierra Club, "National Sprawl Costs Us All," National Sprawl Factsheet, Sierra Club Web site, April 15, 1998.

32. Ibid.

33. Ibid.

34. Ibid.

35. Ibid.

36. Ralph da Costa Nunez, *The New Poverty: Homeless Families in America* (New York: Insight Books, 1996).

Minimum order $10.00 and up

Free second pair

Sizzlin' hot deals!

Free hot fast delivery

No minimums, no monthly fee

Please specify type of payment when ordering: Cash, Visa, Master Card

See reverse side for great savings

Minute after minute your savings multiply

24 hours a day, 7 days a week!

Some restrictions may apply, services not available in all states

Door to door service

Shop online, any time

Overnight delivery!

Printed on recycled paper

For multiple calls, don't hang up. Press #.

Faster, better

Full-size convenience

Price breaker all-in-one!

Quality you can feel

Bulk rate U.S. postage paid

Get peace of mind—extend your warranty

Make low fixed monthly payments

Your brands. Your way. Next day.

Everything you need

CHAPTER FOUR

Possession

Americans Built Their Country Partly of Knickknacks,
Electronic Gadgets, and Paraphernalia

The *Wall Street Journal* once wrote that "how we spend speaks eloquently of who we are. . . . [T]allying cash-register receipts . . . amount[s] to an evolving national portrait: concrete evidence of changing family patterns as well as subtle shifts in our beliefs and priorities."[1] These shifts in our beliefs and priorities may not even be so subtle. They are as large as the differences between slow rural life and fast-paced Wall Street financial brokerages, or between people who read books and people who watch car chases on television. Measuring our cash register receipts as an indicator of how our country is changing offers us a way of looking at America that is different from the ways practiced by our newspapers, politicians, and academics.

We define ourselves through the use of the objects we buy and the services we pay for. The choice of buying a high-rise apartment in the city versus the choice of building a house in the country; the choice of cooking at home versus the preference for eating out. When the objects that we rely upon, like computers and airplanes and umbrellas, break, or do not serve our wishes, we realize how central a role they play in many of our activities, in our work, and in our play. Without our possessions, we would spend our time differently, our meals would take longer to prepare, our travel would be more constrained, our access to information would be hampered, and our expectations for what we should accomplish in a day

U.S. Households Using Various Appliances, 1997

Type of Appliance	Percentage of Households Using the Appliance
Central air conditioning	47.1
Room air conditioning	26.1
Clothes washer	77.4
Clothes dryer	71.1
Dishwasher	50.2
Microwave oven	83
Oven	98.8
Range	99.2
Refrigerator	99.9
Water heater	100
Stereo equipment	68.8
Color TV	98.7
1	31.9
2	37.4
3	19.1
4	7.6
5	2.8
VCR	87.6
Personal computer	35
Cordless phone	61.4
Fax machine	6.2
Answering machine	58.4

Source: U.S. Bureau of the Census

would be different. Imagine trying to start your car in the morning and finding it dead, or going to your computer to write and finding it lifeless. These objects have remade our lives, making our moment in history distant from that of even just a few decades ago, when many people chopped wood and sat at home in front of the fireplace.

The rate at which we adopt such new devices is accelerating. The first mass-produced automobiles chugged down America's dusty roads in the early 1900s, but according to W. Michael Cox and Richard Alm, the country still had more horses than cars into the 1920s. It took fifty-five years to get automobiles to a quarter of the population. But a quarter of U.S. households owned a personal computer within sixteen years of their introduction. The telephone took thirty-five years before it reached one in four homes. But for the cellular telephone, the time took only thirteen years. Electric power existed in the late 1800s, but didn't become nearly universal, even in urban areas, until fifty years later. But the Internet took only about seven years, by one estimate, to reach a quarter of Americans. From airplanes to radios to VCRs, each new invention has spread more quickly than the one that came before it.[2]

The prevalence of the objects that change and shape our lives has changed from a trickle to an avalanche. Many people can no longer imagine how they lived without electronic mail. Many cannot imagine living without several TVs or computers, when one was once enough. Compared to our parents in 1950, people in the United States now own twice as many cars, on average. Since 1960, Americans watch about 40 percent more television, on average. Meanwhile, the quantities of airfreight shipped around the world to meet our appetite for computers, cars, and TVs has risen steadily in response to our material demands. As noted in the previous chapter, we have acquired personal space at a level unrivaled in history.[3]

These changes are not just technological (although technology cer-

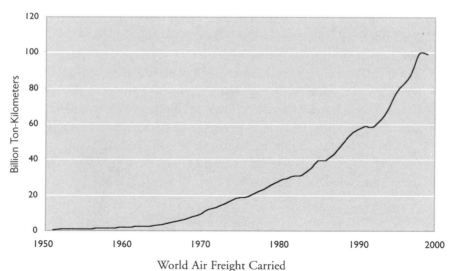

World Air Freight Carried

Source: International Civil Aviation Organization. Figures for 1950–69 exclude former USSR, and 1998 figures are estimates by Worldwatch Institute based on ICAO projections.

tainly plays a large role). The simplest possessions can change people's lives. A tree house where kids go to play can mean a great deal to them. A football and a playing field can be the center of life for some kids. A bed-and-breakfast where people can get away for a day or two can work wonders. A new pair of shoes. A new shirt. All of these can change people's attitudes. Yet most of them do not depend very heavily on technology. Many of them can change people's personal relationships. They can affect people's health. We may regard them publicly as trivial, but people care about them.

For its part, though, technology has changed people through most of history. In 1620, Francis Bacon said that "no empire, no sect, no star," has exerted greater influence over human affairs than a handful of "mechanical discoveries"—namely the printing press, gunpowder, and the magnetic needle (compass).[4] Without tools, our species might never have survived the test of nature or built civilization.[5] Today, new tools are being invented at the fastest rate ever. A computer from last year is already out of date. Safety devices on cars are evolving quickly. Tools for hunting and gardening once shaped the evolution of our civilization and perhaps even our biological evolution as well. Now tools are shaping the evolution of our values by exposing us to longer lives, more travel, more opportunities, more information, television shows, and greater convenience, as well as to new responsibilities.

That these changes stem from plastic and metal and polyester as much as from human values or human wills does nothing to reduce the magnitude and historical uniqueness of these developments. These are the concrete changes in the ways that Americans spend their time, in how frequently we see our friends or neighbors, in how we look at our world. These changes are the evolution of life in America—socially, environmentally, economically, and culturally.

Living with What We Buy

The changes brought by objects are tangible. Yet they have intangible effects that often go unnoticed. Indeed, many of the changing realities of life in America can be traced to people's ownership of comfort-generating and work-saving devices.

Before clothes dryers, people could see their neighbors' laundry drying on the clothesline in the backyard, which converted a little bit of privacy into familiarity. In countries today that lack laundry machines, women (and sometimes men) gather by the river to wash, and to recount stories about the previous day or to share advice. The act of creating the washing machine saved Americans from some toil. But it did more than that. It also took from them one of their opportunities to see other people. It was a victory for privacy but a defeat for socializing. It was a change in society, not one voted for in an election but rather one brought on by two metal devices called the washing machine and the dryer, and yet a change that reached into the details of daily experience and work.

Before refrigerators, shopping was done every day. The same people would see the same shop proprietors, and their fellow shoppers, during the course of daily errands. It created a daily rhythm, a set of interactions, a time when local issues could be discussed, at least in passing. The refrigerator reduced the spoilage of food, made eating safer, and eliminated some of the work involved in shopping and cooking. But it did more than that. It changed social interactions. It even made some families more reclusive, more private, more introverted, since it let them stay home more days of the week.

U.S. household ownership of air conditioning units rose from about 17 percent in 1960 to 50 percent by the early 1970s, and then up to two-thirds of all households by the beginning of the 1990s. This change meant that people were cooler. But perhaps more important, it meant that some people who once sat out in the cooler air of the front porch in the evening probably moved inside.[6] This separated them from their neighbors, and from the sounds and smells of nature. It changed the experience of living in the United States and changed the relationships between people. This would be a different country if air conditioning had not been invented, or had not become popular. It might be a country of stronger neighborhoods, where people would spend more time with their neighbors. It would be a country that burns fewer fossil fuels.

Television is perhaps the object that has most dramatically changed our experience. More than 90 percent of American homes have a color TV today. American children, on average, spend 1,680 minutes per week watching television. They spend an average of 38.5 minutes a week in conversation with their parents. It may be that kids have never talked much with their parents. But the experience of TV viewing is certainly a new influence on childhood.[7] It changes the topics that kids think about. It makes them more sedentary. It changes their relationships with their parents and with other kids. What could be more fundamental?

The average American spends one entire year of his or her life watching TV commercials. American teenagers are typically exposed to 360,000 advertisements by the time they graduate from high school.[8] Whether they are persuaded to buy the things advertised depends on whether advertisers have properly targeted their demographic, but socialization of our children and our adults into a modern consumer culture by means of television is virtually universal.

But the television has other effects as well. In 1980, only about one home in a hundred had a videocassette recorder. Ownership of videocassette recorders rose from only a few households to more than half of all households by the mid-1980s. Many people who went out, either to the movie theater or to other locations, or who read or talked with their families, spent some of their time watching videos instead.[9] The VCR became an inadvertent instrument of social change. In some cases, it may have diverted people's attention from the downtown area of their town to the sights of Hollywood; in other cases it may have taken them away from a game of catch in the street or the park and into the house. These are changes in the family, brought by a possession, and they will surface as well in slightly different values and beliefs among the children who grow up watching those videos.

These unplanned social changes are the continuation of similar, earlier changes that came with washing machines, dryers, and refrigerators. These eliminated some of the day's errands, some of the contact with other people that came with those errands, some of the cherished or unwanted flavors of life. Many of these changes became universal long ago, like ovens and indoor toilets. Some are more recent, but nonetheless prevalent features of our lives. For example, more than two-thirds of all households have a microwave oven.[10]

Computers are already well on their way to becoming dominant features of the American household. Americans already own more computers than any other nation. Of 64 million Internet users worldwide, 41 million are in the United States, and 68 percent of Internet hosts stem from North America.[11] It was U.S. scientists and the U.S. government who invented the computer and the Internet, just as it has often been American individuals and organizations who invented the latest devices. What will this mean for our lives?

Already by the end of the 1990s, for those people who had computers, it meant the ability to telecommute or even to start a home-based business at less cost than in the past. It facilitated electronic trading of stocks and electronic paying of taxes and bills, home video games for kids, access to information about new organizations through Web sites, inexpensive communication with people around the world through e-mail, and, according to a recent Stanford University study, decreased social interaction and longer working hours. People at the firms who manufactured those computers and built those Web sites are especially famous for working long hours, for changing jobs frequently, for sifting through remarkable quantities of information, for dressing informally for work, and for many changes to long-standing traditions.[12]

The changes that accompany all of our possessions, from washing

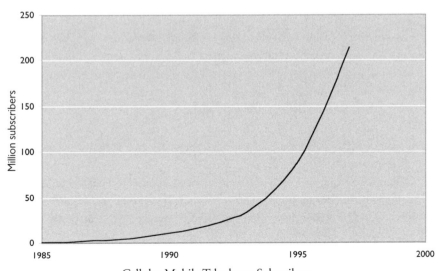

Cellular Mobile Telephone Subscribers
Source: International Telecommunication Union

machines to computers, are far-reaching. They affect our homes, our careers, the ways we spend our time, and the cities that we live in. Historian of technology Barry Katz says, "[T]he design of our cities reflects not conscious decisions about how we ought to live, but arrangements made possible by the automobile, the skyscraper, the telephone; the innermost values of a generation are tied to a world that exists only within the frame of the television screen." The litany of ways that objects, both technological and ordinary, have built our surroundings and changed our habits could fill a book of its own.[13]

Much writing about materialism and consumption has focused on the pollution that comes from manufacturing, consuming, and disposing of those items. But the real effects, including the real environmental effects, are deeper. Consumer objects have shaped the country physically when they shaped our homes and roads and public buildings. But what they changed the most was our values and our activities. They defined who we are as people. It is indirectly, through those changes, that they have affected the environment the most deeply.

The United States is a product of its products, not only through its consumption of those objects and services, but also through the work that it does to make them, the shopping that it does to buy them, and the opportunities created by and the costs of using those items. It is even affected by the disposal of these objects, the regulation of them, the reverence for them, and even the resistance to them. It has been said that America was built by cars for cars. Some people think that the faster our computers become, the faster we need them to be. Others believe that the more we own the more we need.

Producing the United States

When it comes to buying and using products, consumption is one-half of the equation. Production is the other half. For some people, it is more than half. Producing these products often means more for the people who make them, using their sweat and toil, than it does for the people who consume them. Some people measure the success of their lives by the quality of their work and the devices they build. Their lives are shaped by the making of products.

This production of consumables has changed American lives funda-
mentally. Our historical exodus from the farm and entry into the city, for
example, changed both the products that we make and the ways that we
live. This shift changed the ways that people think of themselves as well,
replacing farmers with managers and ranch hands with assembly line
workers. It changed people's status in society. It changed the creases on the
skin of their hands and faces. It made laborers into typists, and turned
back-breaking work into days spent under fluorescent light. In terms of
human experience, the shift that originated with the use of new agricultural machines and factory technologies is as basic as that.

These changes are continuing. Some types of work become available
to more people, while others become rarer or disappear. To keep up with
our shopping, the American workforce is projected by the U.S. Department of Labor to add three-quarters of a million new retail salespersons by
2005. They will offer us the goods—trinkets and cigarettes, cars and
clothes. About 670,000 new cashiers will accept the money as it is handed
over. Roughly 654,000 more general office clerks will do the paperwork.
Due, in part, to the rising accidents and illness from what they offer us
(and to the aging of the population and higher standards for health care),
almost as many new nurses are projected to join the workforce, rehabili-
tating the accident victims, cigarette casualties, and victims of increasing
pollution. Some 648,000 new truck drivers will transport the goods, pre-
sumably also buying as many as 648,000 new trucks along the way.[14]

This is social change measured in the hundreds of thousands of people. The people taking these jobs are also leaving other jobs—entering a
transition from one kind of work and identity to take on another. Their
past jobs may not have been more desirable nor less mundane. But their
new efforts will be directed toward promoting certain products, some of
which help people attain independence, some of which change people's
physical appearance, some of which do nothing in particular. Their status
in society will reflect, in part, the type of work they do and the products
they sell. The ways that they use language will adjust to their work as retail
salespeople or used car dealers. The new truck drivers will spend more
time away from home. The cashiers may find themselves thinking frequently about money. These workers' lives will be shaped by their daily
activities in shops and on roads more than they will be shaped by their

The American Workforce 1992–2005

JOB GROWTH		JOB LOSS	
Occupation	*Increase (No. of Jobs)*	*Occupation*	*Decrease (No. of Jobs)*
Retail Salespersons	786,000	Farmers	−231,000
Registered Nurses	765,000	Sewing-Machine Operators	−162,000
Cashiers	670,000	Cleaners/Domestic Servants	−157,000
General Office Clerks	654,000	Farmworkers	−133,000
Truck Drivers	648,000	Typists & Word Processors	−125,000
Waiters/Waitresses	637,000	Domestic Childcare Workers	−123,000
Nursing Aides & Orderlies	594,000	Computer Operators	−104,000
Janitors & Cleaners	548,000	Packaging Machine Operators	−71,000
Food-Preparation Workers	524,000	Precision Inspectors	−65,000
Systems Analysts	501,000	Switchboard Operators	−51,000
Home Health Aides	479,000	Telephone & Cable TV	−40,000
Secondary-School Teachers	462,000	Installers & Repairers	
Security Guards	408,000	Textile-Machine Operators	−35,000
Marketing/Sales Supervisors	407,000	Bartenders	−32,000
Fast-Food Counter Workers	308,000	Butchers & Meat Cutters	−31,000
		Telephone Operators	−24,000

Source: Harper's Magazine, reprinted projections from the U.S. Department of Labor.

participation in democracy. Their loyalties, the pressures of their days, their identities will revolve, in part, around their work, and the new kinds of shampoo or deodorant that they make or sell.

What are the occupations in which we'll see the most growth? If the Labor Department is right, food service tops the list: By 2005, some 637,000 new waiters and waitresses will serve up hamburgers and fries, and foie gras, and 524,000 "food-preparation workers" will supply the food. Jobs for nurse's aides and orderlies are projected to gain the next largest number of jobs, possibly in part because of the foods delivered by the waiters and waitresses. Some 548,000 new janitors and cleaners will clean up the mess.[15] Of course, all of this takes place in the context of a growing population. These types of production and consumption result in part from our selections at the shopping mall, and in part from the fact that there are more of us all the time. And of course, these jobs are not new. Many of us have always been waiters and nurse's aides and janitors. The large additions to the numbers of us working in these fields are the acceleration of changes that have been taking place for decades in the types of work that we do, as economic growth propels our tastes into an ever larger workforce dedicated to serving them.

The experience of living in America is largely about this evolution in work, the journey away from the farm, into the factory and the office, and onto the Internet. Work is, after all, the way that many of us spend half or more of most days. The kinds of work that we do are making us into different people, affecting the topics that we think about and the ways that we use our bodies. According to one estimate, three-quarters of all U.S. jobs will involve full- or part-time computer operation by the end of the century.[16] Jobs that, historically, required physical work are being replaced by hours in front of a keyboard. We are becoming, as a result, more sedentary.

We are also becoming more technologically literate. Jobs that required conversation now sometimes substitute electronic mail for face-to-face interactions. Work that was done in person is now sometimes done over the telephone. For those people who no longer work face-to-face, the importance of physical appearance is reduced. These are all deep changes in the ways that people live, and in what they worry about.

Shoshana Zuboff, a social psychologist, has explored the unintended consequences of the automation of office work. Today, one clerk standing and asking a question at another's station more often than not signals a problem to supervisors, she says. In the last thirty years, according to Zuboff, the office may have grown quieter, and it has also become more tense and lonely. The kinds of lives that we lead in the workplace are changing. Even our physical health is changing, as our eyes strain to look at computer screens instead of human beings, our forearms ache from carpal tunnel syndrome, our backs suffer from sitting too long.[17] A whole industry has appeared to supply ergonomic office equipment and furniture, from wrist rests to keyboard trays to lumbar cushions. Our personal histories and the history of our country are the cumulative experience of our days at work producing and our evenings at home consuming. Those histories are changing.

Shopping for Our Personality

Making these objects has turned farmers into assembly line workers and ranchers into office executives. Consuming them has changed our homes and our ways of life. In between those two activities of producing and consuming is a third activity—shopping—and it too has shaped our national personality and redefined our experience of public space.

There are more shopping centers in the United States than high schools. The last year when there were more high schools than malls was 1987, and the construction of malls has outpaced the construction of new schools for many years. If prevalence is an indication of importance, it's worth asking which institution has the greater influence on the minds and culture of Americans. The 32,500 shopping centers that were in America by the early 1990s may shape our future as much as education is supposed to shape it. In many community high schools, we only have weathered old books and out-of-date computers. But the mall has the latest new gadgets, the quirky new electronic games, the glossy magazines, the hippest shoes.

Americans spend an average of six hours per week shopping, but less than 40 minutes a week playing with children.[18] Shopping is winning out over kids and over high school by holding more of the attention of our adults and more of the contents of our pocketbooks. Statisticians say that American parents spend 40 percent less time with their children today than they did in 1965. Which activity, then, is shaping our future—family dinners or shopping? The average American adult spends nine times more time shopping than playing with children.[19]

Americans can choose from over 25,000 supermarket items, 200 kinds of cereal, 11,000 magazines. It is no wonder that we spend so much time at the mall. It is no easy task that we face there. We have more choices in the department store than we do in the office. We do not want to squander our family's money. We do not want to wind up wearing an unfashionable or unflattering shirt. Our self-image may be drawn in part from what we wear, particularly for those people who are dissatisfied with their jobs. Shopping is more rewarding to them. Part of why they go to work at all is so that they can increase their purchasing power.

Even those people who spend all of their time taking care of their families often find themselves at the store. They push six, eight, maybe twelve big brown grocery bags to their cars in big shopping carts. For them, the supermarket is much more than a collection of apples, oranges, and household cleaning detergents. It is a social device. The time savings that it offers creates opportunities for them to do other activities, like travel or spend more hours at work. Without the supermarket, we would be tied more closely to the activities of food gathering, cooking, and even farming, gardening, and other domestic chores. With it, we can squeeze in soccer games before dinner and a film after dinner. The supermarket makes it possible for us to arrange our lives and fill them with the activities that we want to do—perhaps we could even spend more time with the kids.

The same is true of the shopping mall. The mall has become a place where people gather. For lack of any other central place to collect, the mall has become a teen hangout. Teenagers eat there, acquire their fashion sense there, see their friends there. Some stores have had to chase teenagers away, sometimes by playing music that young people don't like. When many young people get out of school for the afternoon, they choose to go to the mall—teenagers have responded wholeheartedly to it.

For many people, of all ages, the mall is the place where they do their business, run into their friends by accident, and go to look at other people, see what they are wearing, hear the gossip, judge the latest fashion trends, eat, and meet new people. The mall is about much more than just shopping. It is now what the town square once was. It is where we watch each other. It is where many people go to see what their neighbors like, want, and choose. During the 1980s, 1,500 new shopping centers were built every year, in part because malls fill all of these roles in our society, and in part because the money generated as a result makes their construction possible.[20]

The mall may even be a tool for fighting loneliness. Some people go there to feel a connection to the crowds of people, and to commerce and desire. For people who cannot afford membership at a golf club or other exclusive membership organization, it is a place for social networking. People who do not belong to any social, religious, or other group can belong to the mall and its clientele. Some dates are arranged at the mall; and many of those same dates are held at the mall, with its movie theaters, dinner establishments, and flower shops. Some agreements and business deals are forged there. The mall is a gathering place, one of the few available to some people who may not have very many other places where they can go.

Our personality is shaped through shopping, but not only by what we buy. We shape our personalities through the activity of shopping itself, the pageant of Christmas presents, birthday presents, our platinum credit cards, the wrappers and big bags and bright colors and bows and ribbons, the parking tabs validated at the shop counter, as well as the acquaintances who we run into in the parking lot and the food court. Our high schools can hardly compete with this. When it comes to schools, we have people whose work it is to develop curricula that will take care of our children and our future. But when it comes to malls, market researchers determine how to most effectively move money from the pockets of young people, and people of all ages, into cash registers. Schools do not always succeed in educating us or enhancing our future; but the messages of mall culture make no effort to do so at all—they may even do the opposite. The mall may teach that a successful person is one who wears the right clothes. It teaches that a successful person is one who can afford to buy what he or she wants. It teaches a very different message from those that our schools at least try to teach.

Even Main Street teaches a different message from that of the malls. Main Street in many towns has homeless people. It has the outdoors, with at least a few trees and birds. It does not have Muzak piped into the background. At the mall, security guards can remove the homeless because it is private property. Downtown, the police sometimes do the same, but they are constrained by each person's legal rights to occupy public spaces, and sometimes they are monitored by journalists. Downtown is more real, with real-world problems to see and learn about, from pollution to poverty. Spending the afternoon at the mall leads to a different understanding of the world.

We could develop our national identity through our education system. We could develop it through close relationships with our neighbors and extended family, or time spent in our cities. We could develop it through vigorous physical exercise and our physical well-being. Or it could be through hard work, a puritan work ethic, public commitment, values, or many other ways. But instead, our personalities, both individual and national, are increasingly being shaped by the culture of the mall.

A New Value System

A first-grade teacher once collected old, well-known proverbs. She gave each child in her class the first half of a proverb, and had them come up with the rest. Some of the answers speak to the view of the world that our kids are learning: "You get out of something what you . . . 'See pictured on the box,'" answered one child. "If at first you don't succeed . . . 'Get new batteries,'" said another child. It is not trying hard, but rather an object that can be purchased at the checkout counter that makes success for some youngsters in America today. A third child answered, "A penny saved is . . . 'Not much.'"

Our relationships with objects have changed our values over the years, and our expectations from the objects we buy have changed. Imagine some people from 1776 walking into a home today and trying to identify a thermostat. Someone would have to explain to them that we no longer need to chop wood for our fireplaces to heat our homes. What about a stereo system? Someone would have to tell them to sit and listen, rather than trying to play musical instrument themselves if they wished to hear music. Imagine those people in the kitchen today. What would they make of electric refrigeration? But it's very possible that they might, having understood the utility of these modern amenities, find them capital ideas.

Now, imagine these people walking into the massive New York location of the toy store FAO Schwarz. What would they make of this barrage of toys that children can play with in their rooms? Would they be shocked at the frivolity that had grown from the principles with which they had founded the nation? Or imagine them pondering the awesome varieties of computer configurations in a computer showroom. Would they be impressed or dismayed by the complexity of our technology? Of course, few of our activities today do not use these devices. What would these people from 1776 think of the culture that has grown up around them?

How about the advertising that stares at them from cereal boxes and billboards? Advertising does not tell us that justice will change our lives. It tells us to use objects to change our lives. It does not say "live free or die." It says that hair spray or acne cream is the way to better living. Advertising is far more frequent a message than political dialogue in America. Its greater frequency mirrors the dominant role that products play in our lives, greater than the role that politics plays. We cannot really know what visitors from 1776 would think, or whether they would be too confused to even understand what they would see. But, as hard as it is for us to imagine, we do know that the average colonist in 1776 was more likely to be greeted by a political slogan than by an advertisement for a product. Indeed advertisements for products were virtually nonexistent. Political sloganeering, though, was all the rage, and in all probability reflected the importance of political principles to the people of the time.

Imagine if a politician today could take credit for supplying better mattresses, or discovering aspirin. That would be a powerful campaign slogan—"better sleep for all and headaches for none." But politics cannot provide those. It is industry that can, and so people look to industry for their needs. And industry replies. It answers with advertisements, and then with boxes delivered to the front door. It reaches out through television. The statisticians say that the average American sees more than three thousand ads a day.[21] Industry can afford far more television time than local politicians or civic groups can.

The result is a shift in our principles from the abstract to the mundane, and from public discourse to the marketplace. The result of this shift is that we have ceded the keeping of our principles to economists, and the meaning of value to what can be bought. Many people decide where to live based on the proximity of job opportunities and shops that can offer them what they want, rather than based on long-standing membership in

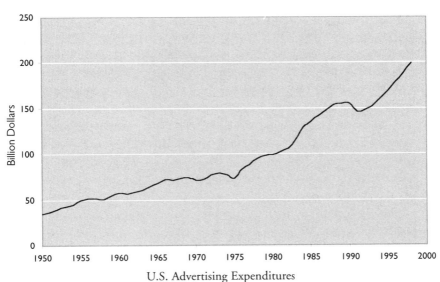

U.S. Advertising Expenditures

Source: Compiled by Worldwatch Institute from Robert J. Coen,
McCann-Erickson, Inc., New York

a community or affinity to a local way of life. The economists' definition of value is heard more and more, from the radio waves to private conversations. Economics holds a larger share of our public space than politics does.

Economic interests control the advertising media, the billboards, the TV, the junk mail, new lingo in shops, and Internet banners. They shape much of our culture, from decisions of what music to produce and market to decisions of which fashions to sell. They even shape much of our politics, where economic growth is the supreme goal. Economics is ascendant. It fills the streets with cars and homes with furniture. Value means purchasing power.

This economic solution is geared to a time when many people cannot agree on common values. Economics is the solution to that lack of agreement. In economics, each person can choose their own path. The marketplace offers a selection of choices—private space, mobility, information technologies, comfort. We can select from those choices in proportion to the amount of purchasing power that each of us has, and we can live according to the value of what we can pay in dollars. Given enough space by a prosperous economy, a wide-enough range of information, enough opportunities and choices for places to live, people can, more or less, live side by side while living according to a variety of values. We need not forge common choices or constrain our individual selections. The marketplace for objects frees us to follow our own ideals, with fast transport taking us to the homes, temples, and occupations of our choices, and all sorts of gadgets making new kinds of work and recreation possible.

People who could not have a dialogue about common choices can avoid common choices entirely and, instead, spend their dollars to build subdivisions where neighbors live far enough apart not to disturb each other. We can buy televisions to entertain ourselves in the evening instead of discussing morals or politics over dinner with colleagues. Monetary wealth, larger houses, and fast transport spread family members apart as well, making us a little more separate from each other, and obscuring whatever disagreements we might have. We buy and sell the commodities, like large houses, fast cars, and solitary entertainment, that make it easier for us to avoid common choices.

Such a process avoids the split between the political left and right. Driving more does not conflict with the policies of either political party. Owning a larger home, a second home, or one's own apartment does not signal a preference for one of those ideologies. Watching television, eating fast food, cooking in a microwave oven, or using any other device of comfort or ease is antagonistic toward neither mainstream ideology of the country. The representatives of both political parties leave these topics out of dispute, possibly because they lack the courage to address these issues, but more likely because they are so much a part of all of those activities that they would never think to dispute this way of life.

In this sense, both parties leave the natural environment out of their ideologies. Prices, the marketplace's version of values, become the guiding principle in the use of nature. The higher a price tag an object carries, the more it is worth; the less expensive an object the more abundantly it will be used and discarded. Natural resources hang in the balance; landfills accept the results; the atmosphere receives the waste gases. Prices lead our decisions, not morals or community.

What Are We Buying?

We are the world's foremost buyers of toiletries, knickknacks, sporting goods, fashion accessories, footwear, auto parts, office supplies, electronic gadgets, junk food, and paraphernalia. Our shops will not be undersold, and many Americans will not be underbought.

What are the messages that our economic behavior sends? In a country where our way of life is determined in part by the ways that people use their money, these decisions are monumental. To the extent that shopping rather than politics determines the experience of living today in North America, important decisions that have at earlier times been reserved for voters and public hearings sometimes now revolve instead over items as seemingly trivial as potato chips and cigarettes and their effects on public health and society.

Junk food, from Fritos to potato chips to Twinkies, is advertised to us on a budget of $366 billion a year, which, by comparison, is more than American military spending. And then, at least in small part because of poor diet, we spend between $33 billion and $50 billion on diet plans to try to lose the weight gained.[22] Remarkably, the world's champion shoppers are spending $44.5 billion a year to buy cigarettes—and then $50 billion a year on health care attributed to illnesses that come from smoking.[23]

Citizens may say in polls that they care a great deal about health care, but they act against it when they base their diets on junk food, cigarettes, and coffee instead of vegetables and fruits. Citizens may vote for environmental protection, but they work against it when they drive too often or select an inefficient vehicle. When the richest people in the world go shopping and choose to buy products that, immediately or eventually, will make them ill, this makes heart disease and other illnesses a larger part of the national life than necessary. Their behavior as consumers contradicts the values that they espouse publicly; and their consumer marketplaces overwhelm their political forums in their ability to determine what our world is like.

It is, in a sense, a form of voting. In a sense, it is every bit as important as any vote made in a ballot box. The connections between our lives and the votes that we make with our dollars, diets, and schedules are much more direct than the connections between our lives and the votes that send representatives to Washington.

A $25,000 sports car may seem an essential purchase for those who want to cruise the strip, in Las Vegas or Washington, D.C. But do sports car owners and other Americans really want to spend $137.5 billion a year nationally on medical insurance payouts and legal costs for the pileups, fender benders, and lethal collisions that happen every year on our streets? Apparently. Spending that cash on insurance costs and repairs is one of the choices that American consumers made this year, last year, and every year before. Public debates and government are hardly involved. This is a form of indulgence carried out between the insurance buyer and the insurance seller, with the federal government's regulators watching from the side. Yet it affects us more than most government policies and more than most topics that we debate publicly.

Would bargain hunters put down the money needed to make better schools, more literate children and adults, and better trained workers if a Madison Avenue advertising firm offered them a cash rebate of $120 billion a year? In a sense, this deal is available, because it is the amount spent by American consumers on crime, for the alarm systems in our homes and the "clubs" on our cars and the costs of prisons. We reject part of this cash rebate when we decide against investments in education and training that could help to reduce crime. Do we want our money back on the fifty-five new prisons that we bought in the last ten years, or the more than 1,500 new prison beds we buy every year, or the total of $31 billion we paid in 1992 to run our prison system? Apparently not.[24]

These problems are sometimes described by economists with words that fail to communicate to the public. The economists talk about such problems with terms like "negative utility" and "externalities." The public might think of them instead as poor choices made by shoppers. Another way to think of them is this: They are part of what happens when we allow economic behavior to substitute for healthy public discourse. We live at a time that has the most advanced products ever found by shoppers or explorers. We have medicines that can fight the scourges that threatened humans through all of history, and yet we spend as much on items and activities that cause illness as we do on health care to fight those illnesses. Then, when we complain about the size of our national health care bill, the political debate focuses on whether the free market or the government can best allocate the payments. We have cars and airplanes that can take us anywhere, but we also have careless driving that contributes to almost 40,000 fatalities, and millions of injuries, every year, and that makes people more sedentary than they need to be. Time and again, the consequences of our purchasing decisions contradict the goals we set for ourselves.

The United States should lead the world in knowing how to use its wealth wisely. Instead, it exports its shopping habits through the packages that it sends abroad, the television shows that depict American choices of what to own, and its participation in international agreements on trade, investment, and other economic ideals.

The Value of Having and Not Having

Alexis de Tocqueville, in *Democracy in America* in 1835, wrote that the loss "of self-restraint at the sight of the new possessions they are about to obtain" would cause Americans to "lose sight of the connection that exists between the private fortune of each and the prosperity of all."[25] He was pointing to changing values as a result of material possessions. If he could see us now, he might remark on the extent of our privacy, our possessions, our individuality.

Much has been written about wealth and materialism, and much of that writing has lamented conspicuous consumption. What many people understand from this is that consuming is wrong, and that it will not meet their emotional needs. People do not receive this message well, however. The reason is that conveniences, opportunities, and laborsaving machines are about more than desire. They are about a way of life. They are about privacy. They are about how many places we get to go in a day and how many people we get to see. They are about personal choices. People resist giving them up because doing so would mean change on a deeper level than just forgoing some gadget. In some cases, it would even mean changing their personalities.

This debate will only become meaningful if it takes place at the level of our identity as a country and our values as individuals. It is not only about how many tons of garbage arrive at the landfill. It is about whether people spend more time in their neighborhoods, or venture farther afield. It is about loyalty and hurriedness and responsibility, not just ownership or consumption. Owning fewer cars may mean traveling less. Owning fewer entertainment machines may mean spending more time talking to family members—which some people do not want to do. Addressing issues of materialism means addressing issues of whom we talk to, how much we talk, what we think about, how much we accomplish in a day, how much we explore our surroundings, how much we stay home, and many other aspects of our personal and professional lives.

These are really issues of who we are, not just our possessions. It is about whether a long hike entertains us, or if we prefer a night of drinking at the bar. It is about whether we want to know more about our local places and nearby people, or about distant places and exotic events. It is about whether we prefer to spend our time talking to people we know or watching strangers on TV. The material results of each of these choices are different.

Our material requirements and demands have changed over time. Earlier in the twentieth century, indoor bathrooms were not taken for granted. People washed their bodies weekly, or not even that often. We may object more to smells today than they did. We take it for granted that we will have music at our fingertips, by turning on the radio. We do not even remember silence, since we have televisions running in the background, or radios, or the hum of engines, all day and all night long. We take it for granted that we should not have to work the soil or raise animals for our food. We take it for granted that we can use birth control. We have fought many illnesses successfully, and many of us now take penicillin, antibiotics, and immunizations for granted as well.

Our material possessions have changed in lockstep with these changing expectations and requirements. Changing what we own and use would also mean changing some of those expectations and requirements, including the ones that define part of who we are as people. Our choices may revolve around objects, be influenced by them, even serve them. But they really come from ourselves and our images of what we want to be like. Without these bathrooms, showers, drugs, radios, and condoms, we would be different people.

These changing assumptions, values, and identities change our world. If we take our identity from what we own and not from the environment we live in, then we will buy and not protect. If we seek to maximize our incomes, then our goals will be the pursuit of GDP, rather than of housing for the homeless or human contact for the isolated or the elderly. If we fashion ourselves after movie stars who smoke, or people who pollute, then we will have that kind of country. We have to choose our values among a sea of material possessions.

People have long debated materialism. The critique of it is an enduring tradition in American scholarship, arts, and letters. Its essential features were in place by the late 1600s, when preachers to the second and third generations of Puritans reproached their congregations for their loss of "heat toward religion" and their "undue affection for the things of this world." Eventually, this religious critique made its way into academia and political debate, such as through the economist Francis Wayland who wrote of the great temptations of "reckless expense," "thoughtless caprice," and "sensual self-indulgence" in 1838.[26]

But these debates have gone by the wayside. By the time a baby born today reaches age 75, he or she will have produced 52 tons of garbage, consumed 43 million gallons of water, and used 3,375 barrels of oil.[27] It is a juggernaut of consumption that seems immune from philosophy. For better or for worse, it is shaping us. The constitution of the United States ensures freedom of assembly and toleration of religion. But perhaps no document can make those ideals into realities today the way that telephones that allow people of similar beliefs to find each other and organize together can, or the way that the cars that allow them to gather can, or the way that many machines can. People using cell phones and all kinds of other gadgets are going to choose what kind of country and what kind of future we will have. Our desires will change the country. We do not, as a rule, consult philosophers or religious leaders before calling on our cell phones or driving to the store.

But if our future is to be determined by anything other than possessions and machines, then our desires will have to include more than the ownership of those devices. This may be happening. It may even be happening on its own, without responding to or acknowledging any critiques of materialism. Our desires may be changing. People once used to "keep up with the Joneses" by buying an expensive car. But many people today are starting to use different measures of success. Some are saying that they would prefer having free time to having an expensive car. Others would rather have autonomy than a large salary. Riches for them are the freedom of not having a boss tell them what to do. They would rather impress their neighbors by being able to take the afternoon off and go fishing. Eight, or ten, or twelve hours in an office following instructions are too high a price to pay, they say, if the benefit is a luxury car or a swimming pool. Today's bragging rights are free time and self-determination.

Those new bragging rights may take our future away from a flood of objects. Telling people not to consume may have no effect on them. But people's own desires for leisure time and autonomy may accomplish what the dire warnings of philosophers never could. Just as objects can affect people's values, so can people's values change their choices of which objects to buy and how many objects to buy. Our recent history has been shaped by the production, consumption, and selection of possessions. But our future could be shaped by different choices. It all depends on how we choose to live in the future.

Notes

1. Christina Duff, "Redefining the Good Life: Indulging in Inconspicuous Consumption," *Wall Street Journal,* April 14, 1997.

2. W. Michael Cox and Richard Alm, "People Want Dead Mice, from the Amazon.com Web site, December 24, 1998, a review of *Myths of Rich and Poor: Why We're Better Off Than We Think* (New York: Basic Books, 2000).

3. New Road Map Foundation, *All-Consuming Passion: Waking up from the American Dream,* 2nd ed. (Seattle: New Road Map Foundation, 1993).

4. Barry M. Katz, *A Historical Romance: Technology and Culture* (Stanford, Calif.: Stanford Alumni Association, 1990).

5. Ibid.

6. Alan Durning, *How Much Is Enough?* (New York: W.W. Norton & Company, 1992).

7. Henry LaBalme, *Fact Sheet,* Washington, D.C.: TV-Free America, 1996.

8. LeBalme, op. cit.

9. Ibid.

10. Ibid.

11. Daniel F. Burton Jr., "The Brave New Wired World," *Foreign Policy,* Spring 1997.

12. James Lardner, "World-Class Workaholics: Are Crazy Hours and Takeout Dinners the Elixir of America's Success?" *U.S. News & World Report,* December 20, 1999.

13. Ibid.

14. "Forecast: The Work Ahead," *Harper's Magazine,* March 1995.

15. Ibid.

16. Edward Tenner, *Why Things Bite Back: Technology and the Revenge of Unintended Consequences* (New York: Vintage Books, 1996).

17. Ibid.

18. Not all of those adults have children, so perhaps the comparison is not fair. But maybe it is fair—overall, the mall is absorbing the time and finances of a larger number of people.

19. New Road Map Foundation, op. cit.

20. "Dodge/DRI Construction and Real Estate Information Service," *Advertising Age,* January 21, 1991.

21. Earl Shorris, *A Nation of Salesmen: The Tyranny of the Market and the Subversion of Culture* (New York: W.W. Norton, 1994).

22. Ibid.

23. Jonathan Rowe and Judith Silverstein, unpublished research, Redefining Progress, San Francisco, California, 1997.

24. Ibid.

25. Mark Sagoff, Introduction to "The Ethics of Consumption," *Report from the Institute for Philosophy and Public Policy,* Special Issue, Fall, 1995, School of Public Affairs, University of Maryland, College Park.

26. Sagoff, op. cit.

27. New Road Map Foundation, op. cit.

CHAPTER FIVE

The Future We
Haven't Talked About

America's Other Public Affairs

Any one of the personal changes in daily life that America has undergone over the past decades would only seem like a curiosity, if examined in isolation. More of us live alone. We travel over a larger distance, on average, every day. We change jobs frequently. These seem like side effects of our economic growth and our wealth. But taken together, these changes in our daily lives have gradually amounted to a revolution. If we look carefully, we can see that they are shaping our lives, and becoming increasingly powerful drivers of our values and beliefs.

Many of our ancestors lived with a lot of small-town snooping, for example. But we have eliminated much of this surveillance. Jonathan Franzen wrote in the *New Yorker* that "In 1890, an American typically lived in a small town under conditions of near-panoptical surveillance. Not only did his every purchase 'register' but it registered in the eyes and in the memory of shopkeepers who knew him, his parents, his wife, and his children. He couldn't so much as walk to the post office without having his movements tracked and analyzed by neighbors."[1]

For many of us, this snooping is nearly gone. Thousands of strangers walk down our city streets, buying what they like and eating what they want—anonymously, either without concern about what the shopkeeper thinks, or at least with no worry that the shopkeeper will tell their families or friends. Most of us probably view this change as a blessing. It is freedom; it is relief; this anonymity may mean more to many of us than the choice of who becomes president, or many other political decisions.

Much of what has changed in American history is personal and subjective, like the freedom from snooping. Some of it has to do with issues like whether we feel embarrassed, rushed, or liberated by the things that we do during a day. These changes may seem less important than large economic or political shifts. But these changes in personal life are every bit as important. They are in many ways independent of the major events of our history, like the rises and falls of the stock market or the elections of presidents. American history also consists of the infinitesimal glances between passersby on the street, peer pressure from neighbors, and tensions between family members, or store clerks and shoppers. Changes in these small, daily interactions are every bit as meaningful to many people as the stuff of state of the union speeches.

It is possible that the United States will continue down a road on which people do less snooping, on which Americans have more anonymity, on which more of us live alone, on which many of us move frequently to new homes. If we continue down that road, then we will have taken an unusual path. We could be one of the few countries to reduce embarrassment, such as that felt by people who buy goods that their families would not approve of. We will be the first whose people spend large amounts of time alone. We will be the first where people can move easily and freely to new parts of the country or to new jobs.

If we continue down the road that we have been following, then where will we end up? Today, Americans travel an average of about twenty-four miles in a day, after increasing this distance continually over the years. Thirty years from now, will we travel forty miles a day on average? If so, then will those large travel distances make us less familiar with the people who live near us, or more anonymous in the places to which we travel? Today, about 10 percent of Americans live alone. Will that rise in the future? Thirty years from now, will 15 percent of us live alone? Twenty percent?

In fact, it is entirely possible that these trends have peaked. Thirty years from now, people may travel the same distance every day that they do now; no more people may live alone than now; people may change jobs with the same frequency as today. These trends could even reverse themselves. The United States could go toward a way of life where people stay near their homes and communities, where most people live in one house for much of their lives, where we belong to tightly knit local organizations made up of people whom we have known our whole lives.

The future of these trends is not clear. Indeed this clarity may be the most glaring absence from the public dialogue in the United States today. It is not clear what direction people want to take. It is not clear whether Americans value stability or constancy in their communities. It is not clear whether we are happy with the relationships between our personal lives and our national politics. It is not clear whether we like our living arrangements. And the reason why it is not clear is that too few of our newspapers provide coverage of these personal changes in how we live. It is not clear because our politics largely ignores them.

A reinvigorated political discourse would capture these private changes. Do Americans want a country with more shopping malls than schools? Do Americans want a country that would be clocked at fifty miles per day if there were a national measure of speed? Do Americans want to live in a place where the average length of stay in a job is one year, or two years? Do we want a set of public policies and economic goals that promote those ways of life? No one knows whether Americans want those things, because they are not the subjects of political campaigns or many newspaper editorials or articles. The change away from neighborhood surveillance of our actions, and toward anonymity, is not debated in Congress. It would sound silly. No public policy language even exists there to discuss such a topic.

It is often said that too few people participate in politics. The reason why they do not is that politics doesn't acknowledge the things that most people care about. It pays a lot of attention to what has happened to our GDP. But it doesn't acknowledge that our country has become more anonymous, faster, more fluid in people's work and locations. Bringing those topics into political discourse would bring more people into it as well.

During the 1998 political campaigns, the beginnings of this enriched political dialogue may actually have begun. Several politicians chose to make the topic of sprawl one of their top issues. Doing so gave them access to voters' emotions over how often they are stuck in traffic; over how far they have to go every day to get to work or to their friends' homes; over the aesthetics of strip malls by the side of the road. New Jersey's governor Christine Todd Whitman, a Republican, even made fighting sprawl her top priority.

These new subjects are going to grow in importance. We will have to incorporate them into our public policies and public affairs. If we understand the issues of our day—speed, living alone, possessions, and our tendency to move on to new places and new jobs (as well as the privacy, anonymity, loneliness, and pressures that result)—then we can choose our future wisely. But these are not the topics that are discussed in Washington. Privacy in political debates means protection from government meddling, rather than loneliness due to increasing amounts of space. Yet more people may face loneliness than government meddling. Speed, if discussed in Washington, means faster fighter-bombers, rather than frequent trips away from our neighborhoods. In Washington, "possessions" mean places like Guam, rather than household appliances. And yet it is with the personal, daily details of life that most people's experience dwells.

We need to understand the changes that have swept our personal experiences in order to understand our history. What have these historical forces made of our lives? Do we walk faster than our ancestors? See more? Expect more? Are we more private? More isolated? More secure? Less secure? These are the topics of a real national dialogue.

Maybe America is having a midlife crisis. It shows the symptoms—it has a desire to keep moving on to new places and new things, a desire to go fast, an appetite for new possessions. Even tattoos are becoming more popular. Maybe we are doing the national equivalent of going out and buying a motorcycle. If so, then perhaps we will eventually progress from this midlife state into a more mature one, just as an individual will get over a midlife crisis.

The question may ultimately be about how much we ourselves have changed. If we are still the same people who once lived in the original thirteen colonies, in our Eastern European and African homelands, in our ancestral Salvadoran and Thai villages, slowly and laboriously doing the day's work, then maybe our world will be a reflection of our age-old desires and needs. But if the changes of our time catch hold and indelibly mark our natures as much as they have changed our possessions and our daily habits, then we may really be headed toward becoming a new kind of people.

What would this new kind of people look like? Could we ever value flexibility and opportunity over family? It is possible that we already do, putting off marriage to late ages, moving out of our family home at earlier ages, getting divorced, living alone. Could Americans come to value free time more than luxury cars? Could we value autonomy—freedom from a boss and from being told what to do—more than a high-paying job? And if so, then would that lead the country toward a more environmentally sound future, with less driving and more time spent with family members? Would more people work at home as consultants or telecommuters? Would people work less? Would the GDP go down—and people's perception of their qualities of life rise at the same time? These may be among the most fundamental, and largest, issues that our country is going to face.

New Measures

The first step toward a dialogue about the personal and tangible changes that have swept America is to find out what has happened so far. Up until now, we have not measured some of the things that are the most important to many Americans, and we have put great value on other topics. We have measured the wrong things.

Every day our newspapers tell us what happened to measures like the Index of Leading Economic Indicators, even though most Americans probably do not know what that index means, or to housing starts, even though they do not say whether the new houses are pretty, ugly, large, small, or identical. They say little about the many personal things that people care about enormously, about such practical details as how far we drove that day or how many people we are meeting for dinner.

The most important things are the hardest to measure. Try to count laughter. Try to measure honesty, or loyalty. Because of this difficulty, our society has settled for measures of subjects that are not the ones that we care the most about. The ambiguous realities of our days defy measure. So we fail to include them in politics, government, public discussion, or indicators of well-being like the GDP.

Such indicators could be used, instead, to paint a clearer picture of our lives. Indicators extend our senses. Data that tell us what is happening to other people, for example, can extend our compassion to include the problems and successes of people who live far away from us. It is like binoculars that allow us to see farther. Data also let us hypothesize about the future. By showing us the direction that many trends have taken up to now, they can let us examine what will happen if those trends continue in the same direction, or if they change direction.[2]

If we use the wrong indicators, though, we will misjudge the future and misunderstand other people. If we fail to monitor trends in what happens to the poorest of us, then we will wrongly view our country as richer than it is; if we fail to measure the need for two-income families, then we will miss a key part of what is happening to our families; if we do not see the environmental results of our economy, then we will misjudge our economy. And many more misunderstandings will stem from a set of poorly conceived measures. The effect of this information will be to lead us astray.

We try to get more of what we measure. When we follow popular measures like the gross domestic product (GDP), we are focusing our attention on money, because GDP is a measure of how much money has changed hands. When we focus on money, we are reinforcing our goal of making more money, rather than offering new goals or possibilities. If we focused instead on laughter, then maybe we would try to make children laugh a little more often and make fewer trips to the office to earn more money. If we focused on honesty, through measures of it that appeared in the newspaper every day and were mentioned on the radio, then we would probably watch ourselves more carefully to see whether we tell the truth. Not only would a little bit of our focus shift from money to honesty, but some professions might do business a little bit differently.

"It is no accident that the social and environmental realms that have suffered such erosion in recent decades are precisely those that our systems of national accounting fail to address," wrote Ted Halstead, who recently founded the New America Foundation in Washington, D.C. "Accounting drives policy, for business and the nation alike," he points out.[3] In part, it is the very failure of our measures and accounting practices that encourages some social and environmental problems. Far from being a neutral taking of stock, accounting is a central input into our decisions and one of the tools that we use to make our future.

We have armies of businesspeople flying from city to city, virtually living in hotels, reading *Consumer Confidence Survey,* the *Producer Price Indexes*, inflation, the S&P 500, reports of what has happened to the GDP, and so forth, and then putting their backs and minds to the goal of "improving" all of those indicators. But little or no stock is taken of what these improvements would mean for our family lives, for equality among people, for the environment, or of what would actually be produced by this commerce. Little stock is taken of whether what is produced would be "goods," or "bads" like cigarettes, hazardous waste, lengthy court trials and lawsuits, additional red tape or bureaucracy, increased traffic, or other unwanted additions to our lives.

Many people have been aware for a long time of our lack of meaningful measures of well-being. Robert Kennedy tackled the problem in a speech in 1968, concentrating on the problems of the GDP as an accounting system for the country. In his words,

> The gross national product does not allow for the health of our children, the quality of their education, or the joy of their play. It does not include the beauty of our poetry or the strength of our marriages, the intelligence of our public debate or the integrity of our public officials. It measures neither our wit nor our courage; neither our wisdom nor our learning; neither our compassion nor our devotion to our country; it measures everything, in short, except that which makes life worthwhile.[4]

If Kennedy is right in defining integrity, learning, and compassion as of utmost importance, and if public policies are to address those subjects, then a new set of indicators will be needed.

An article by Cliff Cobb, Ted Halstead, and Jonathan Rowe extended Kennedy's critique of the GDP. It pointed out that even the accounting system used to calculate the measure is flawed. Putting aside for a moment the absence of some of what Kennedy believed was most important to the country, the GDP also fails to make sense even according to the basic arithmetic that goes into it.

> By the curious standard of the GDP, the nation's eco-
> nomic hero is a terminal cancer patient who is going
> through a costly divorce. The happiest event is an earth-
> quake or a hurricane. The most desirable habitat is a
> multibillion-dollar Superfund site. All these add to the
> GDP, because they cause money to change hands. It is
> as if a business kept a balance sheet by merely adding up
> all "transactions," without distinguishing between
> income and expenses, or between assets and liabilities.[5]

But Cobb, Halstead, and Rowe took the analysis farther. Based on ear-lier research by theologian John Cobb and ecological economist Herman Daly, they created an alternative indicator. The new indicator was not intended to replace the GDP, but rather to demonstrate that it is possible to do better. It took economic production as a base, essentially starting out with the GDP, but then subtracted out many losses that occur in the econ-omy—such as the depletion of natural resources and crime—and added in unpaid work done in the home, and other important parts of our coun-try, to arrive at a more complete calculation. Called the "Genuine Progress Indicator" (GPI), it showed a different picture of how the United States had fared during the last half-century.[6]

The GPI includes the values of both market and nonmarket activity within a single, comprehensive framework, and it has a long-term per-spective that the GDP lacks. Among its many components are: income distribution, value of parenting, value of volunteer work, cost of family breakdown, loss of leisure time, cost of underemployment, cost of com-muting, cost of car accidents, cost of noise pollution, loss of wetlands, loss of farmland, other long-term environmental damage, net capital invest-ment, and other categories. Where the GDP looks only at flows of expen-diture, the GPI takes account of the depletion of natural and social capi-tal. As a result, it gives an indication of whether current economic activity can be sustained over the long term.[7]

The GPI shows how a broader accounting lens gives a different picture of the economy. While the GDP has risen almost continually since World

War II, the GPI has not. The GPI rose until the early 1970s. But since then it has declined. A more inclusive measure suggests that America is not doing as well as we would think if we followed the conventional economic measures. Environmental depletion and social problems pull the GPI down in ways that we might not notice if we did not have such an indicator. More information about the GPI can be obtained easily from the Web site of Redefining Progress: www.rprogress.org.[8]

Similar calculations have been made in many other countries during the past few years. From Australia to Holland, people are trying to get a better, and fuller, picture of their recent history. Occasionally these measures are even printed in newspapers or magazines. When they are, however, it is largely as a curiosity. These measures never appear alongside the traditional core set of indicators that most people monitor daily, like GDP, housing starts, and others.

Fortunately, a movement has sprung up around the United States to gather local people together to try and measure the things that they care about, and then to try to change local government and business and communities based on those measures. It is being called the "community indicators" movement. More than 200 cities and rural communities across the United States have community indicators projects to track local conditions, inform policy choices, build consensus, and promote accountability. Some are mandated by local governments; others are from businesses that want to maintain a healthy community around them; others are created by citizens with no previously existing institutional background. Some of the largest and most successful are in Seattle, Washington; Jacksonville, Florida; and Silicon Valley, California.[9]

Direct changes in local politics and communities have come as a result of some of these projects. In Santa Monica, California, changes have ranged from the planting of more trees to bus schedules that make more sense. The results of each project are specific to the local needs of that place and the initiatives taken by the people who participate. Some of the indicators projects have not managed to make any changes in their communities or local politics. They may have managed to make good measures but not to implement any changes, or in a few cases they may not even have chosen well when they set out to gather their data. But others have led to direct changes in their communities. The data on bus ridership used in Santa Monica, for example, were taken up by the city government and used as the basis for significant changes in bus routes.

Some of these community indicators projects have developed path-breaking new indicators. They have succeeded in a handful of cases at measuring human values that previously had appeared impossible to measure. A few people have tried to count the number of stars visible in the sky on a clear night, for example. Through that calculation, they are trying to measure and grasp such topics as how in touch we are with nature, how much calmness we have in our evenings, how much "light pollution" we endure, and how much development we have pursued.

Another novel indicator is the percentage of the people we pass on the street whom we already know, if we go for a short walk. The larger the percentage, the more connected we are to our community. A larger percentage also suggests that there are more people nearby whom we can go to for help if we need it, and whom we can talk to if we feel like talking. This is a far more personal kind of measure than most of the indicators used by economists and printed in newspapers.

Some people have proposed a different indicator to measure the opposite situation. They say that by counting the number of people who have electric garage door openers we can obtain an indicator of people's separation from their communities, their isolation from people who might help them, and their separation from neighbors with whom they might talk. For the garage door indicator to be meaningful, it requires us to assume that people drive into their garages, close the doors behind them, and then stay home. But many people do travel alone in their cars and then close their doors behind them.

The community indicators project in Seattle has included the number of wild salmon still swimming in local rivers in their report on the city's health. By doing so, they tied environmental health to their assessment of civic and human health. Since the dams that we build for electricity and agriculture, and since our consumption of fish, pollution of waterways, and industrial development all affect the salmon runs, counting the fish gives a measure of our human development. And since many people in Washington have depended on salmon for their livelihoods and their diets, counting the declining numbers of fish also gives us a measure of our predicament.

The set of measures developed by the local indicators movement have counterparts at the state, national, and international levels. In New Jersey, for example, a nonprofit group called New Jersey Future has rallied together people from civic organizations, corporations, academia, and state politics to develop a series of indicators to measure how the state is doing and where it needs help. They covered topics from sprawl to infant mortality to income inequality to pollution. With support from the governor, they have a real chance of moving the state's resources and focus toward the opportunities and problems that it faces and that may have been given too little attention in the past. Their governor even issued an executive order directing each state agency to take steps to improve the state of the indicators chronicled in the New Jersey Future report. City and county agencies are also taking up the report and following suit.

At the international level, various governmental and nonprofit organizations have put together books of indicators of environmental, social, economic, and other topics. The United Nations has made an extensive effort to develop parallel data sets from many countries so that they can be compared to each other and then combined to make global indicators. UNICEF publishes a book each year called *The State of the World's Children* to cover the appropriate indicators. The United Nations Development Programme publishes a unique and innovative report each year called the *Human Development Report,* which is particularly strong on subjects of social well-being. The Worldwatch Institute in Washington, D.C., publishes an annual book called *Vital Signs: The Trends That Are Shaping Our Future,* which covers a range of global statistics, including agriculture, energy, population, transportation, security, and health. The World Resources Institute publishes its *World Resources Report* every other year, with intricate environmental data. Many of these titles are available in good bookstores. Many other reports exist, but take more work to locate.

This effort to improve measurement, both locally and internationally, is not new. The first estimates of national accounts in the Western world were done in England in 1665 by Thomas Petty. He wanted to know how much tax could be paid by the country. In France, a similar measure focused more narrowly on agricultural production. But of course these were very incomplete indicators, and they fall far short of the kind of accurate picture that we can hope to have when we make public policies today.[10]

Clifford Cobb of Redefining Progress traces the use of indicators to judge social conditions back to the 1830s when social reformers in Belgium, France, England, and the United States began using statistics to improve public health and social conditions. Cobb points to the early 1800s as the time when indicators really entered government and social issues. In Europe, physicians and statisticians led the way, he says, by looking for ways to understand epidemics in industrial cities. Census data were collected systematically for the first time during that period, and models were made to show how disease was linked to poverty and social conditions.[11]

In the United States statistics were used to champion prison reform in Philadelphia in the 1810s when reformers produced tables of figures showing the number of people in jail still awaiting trial for each of the five previous years. In 1929, President Hoover gathered a group of scientists to report on current social trends in the hope of making national policies based on the information. In the 1960s, a whole social indicators movement flourished briefly in the United States, and then fell away when many people lost interest, possibly because they could not see direct changes happening as a result of the indicators.[12]

Today, we have much more access to data from which we can make meaningful indicators. Satellites collect it from space. Economic agencies collect it from tax records and other commercial sources. Hospitals collect it from patients. Statisticians like epidemiologists know how to extrapolate from incomplete data to make indicators of larger populations or larger issues. Academics analyze the data. Advertising and marketing firms collect data and use it extensively as well. Insurance firms pore over the statistics about illnesses and accidents. Investors study elaborate data on stock prices and corporate earnings.

In fact, firms who make money from sales and marketing sometimes know more about our national character and habits than the people themselves know about it. An opportunity exists to bring this knowledge into consumers' and voters' decisions, in addition to the decisions of salespeople and politicians. We have a great start when it comes to making indicators of how we are doing. But we have a long way to go if we are to represent the complex and changing ways of public and private life in America.

America's Other Public Affairs

Our politics are about problems. We have politics to deal with the issues that confront us: unemployment, poverty, racism, health care costs, social security, foreign policy. Politics is a way of resolving our differences.

But many of the most dynamic issues of our day cannot be viewed entirely as problems. For our lives to speed up as we cover more miles in a day, produce more gadgets with our assembly lines, and do more calculations on our computers is not exactly good or bad. Likewise, our increasing tendencies to live alone and to have smaller households, to pay for more privacy, and to separate ourselves from others are not necessarily mistakes. Many people do not view it as a problem that Americans move frequently to new homes, new states, new jobs, and away from their friends. Yet all of these issues matter to what our country becomes.

When we debate many issues in politics, we call them "tax policies," "subsidies," "housing credits," "budgets," "zoning codes," "growth," and so forth. But when we discuss many of these same topics in private, we call them "speed," "privacy," "loneliness," "traffic," and "work." When we treat them as public policy, people can take sides, and can lobby for or vote for or against a new zoning code or a new tax increase. In public policy, lines can be clear. Some people are for; others are against.

But in our homes, it is much harder to be clearly for or against an aspect of how we live. Privacy is a luxury; isolation can be painful—and yet the living situations that give privacy can cause isolation. Speed gives us the opportunity to do many things, to make money, to see new places. But it also makes some people feel rushed, makes them wish that they had more time at home or with their families, leaves them permanently in need of sleep. The public policies for transportation, housing, and economics that were voted for by legislatures with a clear yes or no have become ambiguous and gray once they reach our homes.

Congress does not usually write bills with the goal of changing the size of our households or affecting how often people leave their jobs. It is not for our government to tell people how fast or how slow to run their lives, even though our society is shaped by the result. Government does, however, take actions that affect the many personal aspects of our lives— but it does so based much more on goals of economic growth than on what those actions mean for the lives of Americans. Government financing of roads fuels part of our speed, for example. But debates in Congress over how much money to put toward roads usually do not raise issues of how much time Americans spend in their own communities and how much time they spend driving away from their neighborhoods. Government subsidies for oil exploration and government taxes on gasoline change how much we can afford to drive. But these are usually treated as issues of economic growth, rather than as topics that contribute to a way of life.

It is time that these debates let go of their central focus on economics and begin to encompass the fuller set of results that roads, tax codes, subsidies, and many other governmental choices have on American communities and lives. Regulation touches myriad aspects of our transportation, our manufacturing, our housing—even so deeply as to affect how many people we live among at home, where we choose to work, whether we move away. The United States is a wealthy enough country to be able to balance its quest for economic growth with its priorities of having people who live well and who live according to their values.

To face up to these issues, we will need a dialogue that can host ambiguity and deal maturely with trade-offs. Rather than requiring a more difficult dialogue, though, this new one may be simpler. Politicians have developed a vocabulary of their own, punctuated by jargon and large institutional bureaucracies and bills labeled by acronyms like the WTO, HUD, DOE, HR1411. Instead, the terms that describe American lifestyles are much simpler ones. They revolve around such desires as adequate parking spaces, free-flowing traffic, cable TV, a good doctor, well-trained teachers, and safe places to live. Politics has become bogged down in old antagonisms. It is combative and it takes sides. The search for a national identity will now require a new political dialogue, one that is tied more closely to daily life.

Our politics may be beginning to move toward the simple things that people care about, like sprawl. "Sprawl is about to explode onto the American scene—moving up in the political agenda," asserts Richard Moe, president of the National Trust for Historic Preservation. If so, then this will be a movement that brings the political debate closer to the frustrations and goals of many people's daily lives. Sprawl is an example of a public issue that is also a proxy for a set of private frustrations. It will tie together some of the changes that are taking place in our neighborhoods, our commutes, and our families with the political debates that have often missed those issues.

Sprawl may be the first subject to cause it to happen. But others will follow. This would be a healing of our political dialogue. It would mend the disconnection that we have had between political and economic jargon, like taxes and subsidies whose merits are economic, and our lives at home, whose merits are social and personal. It would force economists and politicians to accept, and discuss more often and more openly, the social and familial results of economic policies.

The Invisible Strings

One of the most remarkable aspects about economic changes and public policies is that they can change our personal lives in ways that do not have to be discussed during the formation of those policies. Seemingly arcane topics like tax policy can affect where we go during a day, or how much time we spend with our friends. Incomprehensible quantities of money spent for subsidies on oil development or on ranching can affect how much we drive, how spread-out our communities are, how many new sprawling subdivisions are built, and many other topics that we live with every day. But many of those policies are made with economic growth, or other goals not related to the way we experience our lives, in mind.

One of the reasons why people move around so much is that we have housing markets and public policies that encourage mobility. In high mobility countries, like the United States and Australia, relatively low costs of land and building materials and less stringent government controls on building, land use, and real estate markets make it easier for individuals to move.[13] Access to credit is crucial as well. And so housing markets change our families and our selves—without a public dialogue.

They are like invisible strings that change our daily lives by affecting the prices of land and homes and cars while we are not watching. The U.S. government actually pays to increase the congestion from suburban sprawl, for example, when it subsidizes new roads in places that were previously rural by not requiring developers to pay full costs and when it gives tax breaks for the construction of new subdivisions and strip malls. The lion's share of a 1998 congressional transport negotiation of about $217 billion went for highways, for example, many of which will open up new land for development. The more spread-out developments that result will lead to more suburban sprawl. Funding could have gone for a more centralized form of development, with town centers and mass transit. But it did not.[14]

According to the U.S. Department of Transportation, tolls, gasoline taxes, and other user fees cover about 70 percent of the cash costs of building and maintaining the country's road system. The rest—tens of billions of dollars per year—is financed with general tax revenues that we all pay. This means that our taxes fund sprawl and, through it, traffic congestion and a decentralized way of life that polls repeatedly show Americans do not want. Then the government pays again, when it pays the price of air pollution produced by the cars, of mending people injured in accidents, and of other "external costs." In total, the bill for this indirect spending runs, by conservative estimates, to at least 22 cents for each mile driven.[15]

James MacKenzie of World Resources Institute says that for every 10 percent drop in gasoline prices there is a 1 percent increase in domestic travel. This means that our travel schedules, and our interactions with the people whom we visit, respond not only to our own choices, but also to a pricing system that is outside of our personal control. It also means that the cleanliness and stability of our environment rise and fall in part according to gas prices, since every 15 gallons of gas burned in a car releases about 300 pounds of carbon dioxide into the air.[16]

Meanwhile, when we have low-density tract developments, the cost of most government services goes up. Sewer lines are longer, school buses travel farther, more fire stations are needed, and more miles of road must be built, along with new schools and other capital projects that would not be needed if people lived in more compact ways. In Maine, for example, the state spent more than $338 million between 1970 and 1995 to build new schools in fast-growing suburban towns even though the number of public school students in the state declined by 27,000. The same has happened with sewer systems in Tallahassee, Florida. The states are paying more as people respond to cheap roads and spread-out ways of life that the state and federal governments themselves have subsidized.[17]

Subsidies

Subsidies for private transportation and energy use, as well as other subsidies like those for logging, mining, and ranching, may not—often do not—meet the country's expressed goals. A national debate is unfolding over whether we have too much suburban sprawl and whether we spend too much time driving and in traffic. But meanwhile, every year, our government offers a subsidy of tax breaks and road spending, over and above the amounts that drivers pay in fuel taxes and other fees, to make driving possible and easy both on existing roads and on new roads. This has the effect of encouraging low-density, car-based land use patterns even at a time when ballot initiatives around the country have suggested that voters want to reduce our automobile dependency. Some estimates of the amount of that national subsidy even put it at $360 billion a year.[18]

These subsidies are one part of the reason why the United States has by far the highest number of vehicles per person in the world. Norman Myers reports that we have enough cars that every one of us could sit in a car at the same time and nobody would need to be in the back seat—and the majority of those cars would still only have one person in them. Our roads cover 2 percent of the country's land, more than is occupied by housing, and our roads have an aggregate expanse larger than Florida. Roads, parking spaces, and fuel stations cover between one-third and one-half of the total space in American cities. In Los Angeles, it is two-thirds. And Americans drive more miles per year than all the world's other drivers put together.[19]

U.S. energy subsidies are estimated at about $32 billion.[20] This is money paid by the government that makes it much more affordable for consumers to heat large homes, to carry electricity bills on the salary of one person who lives alone, and to afford the price of frequent trips on airplanes. The government is picking up part of the tab for the energy used in each case. It is government support of a spread-out way of living, with fewer people per household than in countries where energy costs are more prohibitive. It is clear that voters want cheap energy; it is not clear, though, that voters want the social and environmental results of cheap energy—it is not clear because no dialogue has occurred.

These subsidies are not only an economic issue. They also affect the sizes of our households, the ways we spend our weekends, the quality of our natural environment, and many other topics. No one at the U.S. Department of Energy may consider those issues when they contribute to energy policy, but they are affecting all of our personal lives just the same. Economists trained in our universities know how to make graphs of the costs and availability of energy, but most of them are not trained in how to make graphs that tie energy policy and prices to nature, to climate change, to social conditions, or to the ways that we spend our weekends.

Water subsidies in the western states are estimated to amount to $4.4 billion per year.[21] This water is used mostly to grow crops, many of them in dry places where agriculture would not be possible, and where few people would live, without inexpensive water. Parts of the populations of Arizona, New Mexico, California, and other places, can live where they do only because of these cheap prices. Major environmental disruptions, from large dams to rivers that no longer reach the ocean to the extinction of fish and aquatic species from lack of water, are all also made possible by these subsidies.

Similar subsidies exist for logging, mining, grazing, and many other activities that are carried on even though they would not be cost-effective without government subsidies and even though they place a large burden on the environment. Citing figures in British pounds, Norman Myers writes, "Can you believe that you pay 2,000 [British] pounds per year in taxes to degrade the environment and undermine the economy simultaneously—and then pay another 2,000 pounds to clear up the environmental damage, plus economic costs such as higher food prices?"[22]

In other countries, these subsidies are similarly large, and sometimes even larger. Myers estimates the global subsidy for road transportation at $917 billion. He puts the world's fossil fuels and nuclear energy subsidies at $145 billion. For water use, it is $235 billion, he says, and for agriculture, $575 billion. To promote fishing on already overharvested fisheries, the global subsidy is $22 billion. Those are the magnitudes of the invisible strings that affect the health of our oceans, pasturelands, and atmosphere, and which affect what many of the people in the world eat, how much they travel, where they live, and many other issues.[23]

Taxes

Those are the magnitudes of the subsides. But the effects of America's tax code, and the tax codes of other countries, are just as large and just as deep. According to Brian Dunkiel, Jeff Hamond, and Jim Motovalli, "Our current internal revenue tax code dates back to colonial times and reflects colonial attitudes." Since then, say those authors, the tax code has increased in complexity, not changed in outlook. When our nation was young, the emphasis was on opening up what seemed like a limitless wilderness, and little thought was given to natural resources. To colonize the "wilderness," tax policies gave breaks to the taming of nature, to logging, ranching, farming, and mining, to the building of roads, and the claiming of land. Despite our growing awareness of the long-term costs of environmental degradation, tax priorities haven't changed.[24]

Federal, state, and local governments combined collect over $2 trillion from us each year.[25] If the force of this spending worked to discourage activities that Americans say they don't want—like pollution—then it would be far more powerful than any other initiative to avoid or clean up after pollution or almost any other problem. But if the effect of this spending is to discourage things we want—like work—and its effect is not to avoid problems, then its potential is lost. It can even cause harm.

Our tax code is never neutral. It is not the case today that taxes neither protect nor harm our environment and economy. Currently, our system of taxation mostly uses income taxes and payroll taxes, and so it taxes our work. When it distributes the tax burden in that way, the tax code takes away some of our incentive to work and so discourages our work, and it also reduces the incentive for employers to hire more workers. These are losses, both for the economy as a whole and for each of us individually. Meanwhile, those same taxes do nothing about the depletion of our natural resources. It would make more sense to discourage those things that we do not want, like pollution and resource depletion, than to discourage productive work.

Taxes don't affect only our pocketbooks. They also affect what kind of houses or apartments we live in, what we eat, and where we work. Taxes affect us as much, maybe more, than safety regulations do. But even though many people take it for granted that we should have a set of regulations that protect our health and our environment and reflect our values, we usually ignore the ability of our tax code to accomplish those same goals. Taxes may be more effective at accomplishing those goals than regulations are, if the taxes are used well. Taxes can discourage dangerous activities, encourage careful safety practices, and encourage the use of clean production processes—provided that taxes are directed toward those goals.

But today taxes are not usually used in those ways. "Some tax shelters hit close to home," according to Professor Oliver Houck of Tulane University. He points out that the tax provision that allows Americans to deduct mortgage interest paid on second homes is a major impediment to the protection of threatened and endangered species. It encourages people to build vacation homes in pristine areas and to build large homes with large yards in places that require long drives to reach. The $43 billion that it cost the U.S. Treasury in 1998 made it the largest U.S. tax break for development. It went, in significant part, to reducing America's natural habitats and to building sprawl.[26]

Many people think of large subsidies that go to polluters as going to big corporations. But during the 1990s, soccer moms and other sport-utility vehicle owners received breaks from the "sport-utility vehicle exemption" from the gas-guzzler tax and were exempt from the fuel efficiency standards of smaller vehicles. This amounted to as much as $7,700 of savings over the life of a Lincoln Navigator. This tax break, which has been received more by wealthy people than by low-income families, encouraged the purchase of large vehicles and exempted those vehicles from meeting ordinary fuel efficiency standards. Sport-utility vehicles, including light trucks, now account for more than half of all vehicles sold, and they put 30 percent more carbon monoxide and hydrocarbons and 75 percent more nitrous oxides into the atmosphere than ordinary cars. The ownership of sport-utility vehicles is also encouraged by low taxes on gasoline.[27]

Our tax code has many surprises. The income tax paid by individuals makes up 30 percent of total tax collections, while corporate income taxes only provide 6.3 percent—just more than a fifth that paid by individuals. Payroll taxes, which cover social security payments, workers compensation costs, and funds for other social goals, provide 28.1 percent, bringing the tax on work to almost two-thirds of tax revenue. But little or no tax is paid on pollution. As Friends of the Earth (FOE), a nonprofit environmental organization, says, "earning an income is something we all like, and pollution is something we all hate, so why do we tax income heavily and overlook pollution as a target for taxation?"[28]

Moreover, according to Hanno Beck, Brian Dunkiel, and Gawain Kripke, some tax laws even give extra assistance to people who deplete or pollute natural resources. A special income tax deduction called a "percentage depletion allowance" encourages the mining of minerals, including toxic ones like mercury and asbestos. A special "passive loss" tax shelter has been created to help investments in oil and gas, but no similar shelter has been created to help investments in cleaner energy or energy efficiency. The bonds used to finance the construction of solid waste incinerators that release toxic substances into the air are exempt from federal income tax. These are just a few of the tax laws that are set up in ways that encourage rather than discourage pollution.[29]

A report from FOE in 1995 identified more than $22 billion in tax breaks for polluters. The report is called *Dirty Little Secrets*. Not only could that money have been spent in other ways if it had been raised, but more importantly, the activities that caused that pollution could have been discouraged. Shifting taxes from labor to such pollution would thus have a double effect: It would both save the money and reduce the harm of pollution, and in turn that reduced harm would save money as well. FOE even found that "under current law, polluters who cause environmental harm can fully deduct all the costs related to illegally released pollution including cleanup costs, legal costs, court settlements, even the cost of the polluting substance itself." So, for example, according to FOE, when the Exxon Valdez oil tanker sunk, Exxon was able to cut its taxes by $300 million.[30]

Proponents of shifting taxes away from work and onto waste and pollution do not mean that all taxes should be shifted. Alan Thein Durning and Yoram Bauman, for example, propose a shift of 10 percent of the federal tax burden over the next 10 to 20 years. They and some of their colleagues propose to purge the tax code of regulations and loopholes that encourage environmental degradation, such as the $17 billion cost of tax-free parking. Much of the shift, therefore, would come from reducing current tax breaks rather than from implementing new taxes. New levies would be applied, however, on pollution-generators like products containing lead, gas-guzzling cars, ozone-depleting chemicals, and the burning of fossil fuels. Taxes would be judged on their real contribution to the economy, in terms of job creation and productivity growth, equity for the people paying them, and resource conservation.[31]

Some states have been using this strategy. Washington State wanted to reduce its road building costs. It did so by giving credits to employers who then passed on the savings to their employees who participated in ride-sharing programs. Nearly $1 million in credits was claimed by commuters, with 55,000 people participating. The state government saved money because it curbed the need for building additional road capacity by an amount larger than what it spent on the credits. It also reduced the number of cars on the roads and so reduced traffic and pollution.[32]

Detailed proposals for this kind of tax shifting have been developed for Vermont, New York, and Minnesota. Indiana taxes the storage of petroleum products. Massachusetts levies a landfill tax of $1 per ton of solid waste added. In Washington, a fee of $1 per new tire sold goes to tire recycling programs. These new types of taxes are working. And they are bringing in revenue that can give the states the ability to reduce their income and payroll taxes.[33]

It is really in other countries, though, where ecological tax shifting has taken hold. In 1991, Sweden shifted 1.9 percent of its taxes away from personal income and onto carbon and sulfur emissions instead. In 1994, Denmark shifted 2.5 percent of its tax revenue from personal income to motor fuel, coal, electricity, water sales, waste incineration, landfilling, and motor vehicle ownership. In 1995, Spain shifted 0.2 percent of tax revenues off of wages and onto motor fuel sales. In 1996, Denmark shifted a further 0.5 percent of tax revenues off of wages and agricultural property and onto carbon emissions, pesticides, chlorinated solvent, and battery sales. The Netherlands moved 0.8 percent of its tax revenues from personal income and wages onto natural gas and electricity sales in 1996. The United Kingdom moved 0.2 percent from wages to landfilling in 1996–97. Finland moved 0.5 percent from personal income and wages to energy sales and landfilling in 1996–97 as well. These are among the examples of successful tax shifting, and they demonstrate that Europe is ahead of the United States on this policy initiative. When the United States catches up, it will have gone far toward cleaning up its economic system, and will have rationalized its social and environmental goals at the same time.[34]

These changes are not intended to affect the distribution of the tax burden. In any case where they detract from the current tax system's progressive nature, in which the rich are intended to pay more in taxes than the poor, then adjustments would have to be made to preserve the progressive elements of the tax code. Overall, in the words of Jeff Hamond of Redefining Progress, "This approach would reduce current taxes on labor, innovation, and capital formation and replace the revenue with new levies on pollution and waste. Total federal revenue would be unchanged, and the current distribution of the tax burden across income groups would be preserved."[35]

The political struggle over such proposals to change taxes and subsidies would be monumental. These are the most fiercely fought politics of all. But consider the reason. They are fiercely fought because they matter. Taxes and subsidies are among the deepest and most effective tools that we have for making public policy. If they had less consequence then it would be easier to change them in Congress. The fact that these politics are so fierce is a strong indication that these issues are critical ones on which to focus. The vehemence with which they are fought matches the stakes that are at hand. If we pass over these issues because of the difficulty in making them happen, then we will pass over the most effective means of building the kind of country that we want.

Pulling the Strings

Unfortunately, most people are not familiar with the details of taxes and subsidies. Economists, tax experts, and politicians make decisions about them, often without considering at all what these issues will mean to other people's daily routines or for the environment. It is only when the public begins to know more about these issues, to hold their representatives accountable for these issues, and to vote on them, that we will be able to improve the way we use our "invisible strings." Friends of the Earth's first recommendation for actions that individuals can take about these issues is the following: "Don't leave the field of taxes and economics to the experts."

Even improvements in public works can affect personal aspects of our daily lives. In the mid-'80s, for example, congestion became close to unbearable on Maryland highway 270, which runs northwesterly from the Capital Beltway into prosperous suburban Montgomery County. So the county and state governments paid $200 million to expand the road up to twelve lanes.[36] But fewer than eight years after the expansion was completed, the highway was again reduced to what one official described to the *Washington Post* as "a rolling parking lot." The daily auto and truck usage was running as high as 210,000 vehicles a day, beyond the official capacity of 190,000—in fact more than state highway officials had projected for 2010. The goal of reducing congestion had failed.[37]

Many transportation experts are now calling this "induced traffic." They say that wider roads attract drivers from other roads, lead people to build more suburbs and homes, and attract people from the city to move out to where they need the wider roads. The effect is to replace the congestion. *Washington Post* writer Neal Peirce says that "[i]n the five years before I-270 got widened, 1,745 new homes were approved in the 12 miles north of Rockville, the major community on the route. In the following five years, 13,642 were approved."[38]

The same has happened in other places. In California, University of California researchers checked 30 urban counties from 1973 to 1990 and found that every 10 percent increase in new lane-miles generates a 9 percent increase in traffic. According to David Walters, transportation expert at the University of North Carolina–Charlotte, "The availability of the transportation acts as a catalyst for more movement, so that the more roads we build, the more places we can drive, the more we drive."[39] In Peirce's view, "[G]overnment actually pushes sprawling development, siphoning growth and vitality from existing cities and closer-in suburbs. City and established suburb residents pay most of the bill. Longer commutes and trips end up generating more congestion, more energy consumption, more pollution."[40]

The reverse seems to work as well. A British team even found, based on analysis of 60 cases worldwide, that where roads have actually been closed, or their capacity severely reduced, an average of 20 percent and as much as 60 percent of the former traffic disappears entirely. Likewise, after a portion of Manhattan's West Side Highway collapsed in 1973, forcing closure of most of the route, 53 percent of the prior trips simply disappeared. "No one can argue New York was seriously damaged," says Peirce.[41]

Many kinds of invisible strings affect our lives. International trade bodies, like the World Trade Organization (WTO), change our lives right inside our own homes. The increased trade and competition that they try to foster is intended to have the effect of bringing down the price of consumer goods, among other goals. These less expensive items mean that we can afford to buy more objects, in some cases meaning that those of us who already own many objects can buy more. We can afford new cars to make it more comfortable to drive often. We can afford airplane tickets. We can buy larger homes. We can buy all of the accessories needed to stock those homes and cars.

In order to get those prices down, the WTO works to reduce or eliminate domestic regulations in the United States and other countries that could be barriers to trade, even though some of these regulations and standards are designed to protect the environment and consumers' health. So this international body can reach into domestic policy and into our neighborhoods. But again, as with many economic issues, most citizens are not familiar with the details of these international agreements, and information on exactly what changes they will lead to in our communities and families is actually very difficult to find. A few studies have focused on the effects of trade on the environment, though that research is still in early stages; no one has come to any conclusions about the effects of trade on families, or on many personal aspects of our lives. No government official has the job of analyzing the effects of changes at the WTO on particular local communities. Few journalists have the specialized experience to report on such detailed changes. In many cases, it would not even be possible to predict what the results of these trade agreements would be.

Many changes in our lives start out as international agreements for trade or investment, or domestic tax policy or subsidies. But what they become in our lives are increased mobility, additional living space, anonymity, and the feelings that accompany those changes. The WTO and the U.S. tax code pull the strings that manipulate our daily routines. We will have to start taking those effects into account when we support or fight against federal, international, and local policies. You can't bump your head on a law or an international agreement. But people in countries with growing fleets of cars can bump their heads during a car accident, or develop a headache or other health problems from the pollution released from the new kinds of manufacturing taking place in their countries. Laws and international agreements may seem abstract, but they manifest themselves in physical ways in our lives, in the lives of people in other countries, and in our environment.

Even highly technical and economic topics, like the setting of interest rates, affect our activities deeply. Lower interest rates promote the borrowing of money, the purchasing of new business capital, the creation of new jobs. This changes our daily routines, as we leave old jobs and take those new ones. It changes our possessions as we can afford to buy new necessities and new toys. The setting of exchange rates has similar effects. If the dollar rises, then foreign goods seem more affordable to us, and we can buy more of them. But a rising dollar makes our goods seem more expensive to people in other countries, so they buy fewer of our goods. This hurts our competitiveness and our export businesses, which reduces the number of jobs. As a result, it changes some of our daily schedules once more, shifting jobs from domestic service work to manufacturing for export, for example, and affects where we live, how much we earn, and what we can afford to own. These economic instruments are not in any way confined to the world of economics. They are a major presence in our homes, even though we cannot see them with our eyes.

Maybe these changes even reach right into our own personalities. Is it possible that we walk faster when we are busier? If we work more hours because of a booming economy, then does it change our pace of life? Does it make us hurried? Or does it make us calmer because we know we can count on having enough income? Do Americans walk faster than people who live in the Caribbean or in India, and if so then is it because of our economy? These questions are probably not answerable, and each of us might have a different answer according to our individual circumstances and personalities. But they are important questions to ask. The results of economic changes can reach right inside of us.

A New Politics

The first step in addressing these issues is talking about them. Sometimes it is said that there are no new subjects in politics, and that we can only revisit age-old debates. But that is not the case in America today. No country has ever debated the merits and costs of having one-quarter of its households consist of one person living alone. No country has ever debated the values and consequences of having a population that moves as frequently to new homes and new jobs as ours does. No country has ever dealt with the environmental consequences of a population that owns as many consumer durables, electronic goods, large homes, cars, or airplane tickets. All of these issues will require new dialogues.

Economists often use the word "development" to mean the building of new factories, dams, and roads. But development really means more than that. It is social change. It is about how we live. Even though most economics textbooks have not admitted it, development today is, in significant part, about topics like speed, privacy, and connection to other people. We can see these changes across the United States, and we can hear the dialogues that are sprouting to discuss these changes.

Julia Heath, an economics professor at the University of Memphis, tells the story of one of these changes. As the commercial market grew in the United States, she says, many of the services that family members provided for each other were replaced by services done for money by people outside the family. Haircuts, for example, moved out of the kitchen and into salons. Teaching, once done by parents, is now the career of professionals.[42]

Heath says that "[a]ll facets of family life have been marketized, or commodified, as the market's encroachment into family life has progressed virtually unabated." Yet little public debate took place about the merits of this change. According to Heath, "The family is increasingly a conduit for market forces, struggling to simultaneously remain economically viable and to weave our social fabric." But Congress has not debated this change. This dialogue will have to grow in the future.[43]

Likewise, without a public debate, our personal and national identities are being formed in part by consumer products that we buy. Political scientist Maryann Cusimano says that many of our corporations would like us to form our personal identities based on which clothes, shoes, cars, and other consumer products we own—you are what you wear, what you consume. Rather than drawing our identities from our families or friends, many of us look to possessions. "Advertisers spend billions to imprint brand loyalty at an early age, and all the advertising of Planet Nike, I'd-like-to-buy-the-world-a-Coke, and Microsoft's one world share a common theme, that identity stems not from national borders but from consumer products," says Cusimano. In this marketplace for identity, we are not born with our personalities, but can buy them and change them with a trip to the Gap or Benetton. No personal experiences need take place. No family meeting need be held to decide the new personality. It can be paid for at the cash register.[44]

We have remarkably few forums for discussion of issues such as the ones that Cusimano is describing, or for discussing the many changes in the ways that we live, or of our changing relationship with our economy. The community indicators movement is one such forum, especially in those communities where researchers develop indicators that include personal changes in how we live. But it does not reach most Americans. The mainstream news media do write occasional stories about Americans feeling rushed, or owning more merchandise. But those stories appear only a handful of times per year, and they do not go into very much detail. This means that our largest and best-funded publications and meeting places are not offering us very many opportunities to talk about topics like our tendencies to live in smaller households or by ourselves, to change jobs frequently, to move to new homes often, or many other related topics.

A few small movements have developed to try to inject new values into our lives or our public affairs, and these movements do host dialogues about our ways of life and our well-being. One such movement is for "voluntary simplicity," and its most notable proponent is Duane Elgin, who published a book by that title in 1981.[45] Thousands of people around the country subscribe to this belief that their lives are richer and happier when they own fewer objects, free themselves from the pressure of trying to earn large amounts of money, and break away from society's goals of economic growth and prolific shopping. These people have a rich dialogue among themselves about those issues. But again, most Americans are not aware of this alternative and are not making any effort to discover its ideas.

Another movement that touches on the issues of how we relate to the economy and to changing conditions in America is the movement for "going local." It is well represented by a recent book by Michael Shuman called, predictably, *Going Local: Creating Self-Reliant Communities in a Global Age.*[46] Shuman and his many colleagues believe that a healthy economy must be in large part a local economy, one that includes products made in the styles of local cultures, from materials taken from the local environment by people familiar with their own region's ecology and history. Buying locally made products keeps money in the community, builds relationships between consumers and producers, allows consumers to see the faces of the people with whom they do business, and maintains the sense of uniqueness still enjoyed by many local communities. At a time of identical, cookie-cutter chain stores and strip malls that look similar no matter where they are located, the idea of local economics and individuality does have appeal for many people.

Shuman has been traveling the country giving presentations about the many success stories that have come from local enterprises. There is nothing abstract about the idea of going local—throughout most of history, most businesses, nonprofit organizations, and institutions were local. Today, many enterprises that stay in their communities have brought more jobs to those places than multinational corporations have, and the local companies have left the profits in those communities. Shuman tells success stories of inner-city areas that are now showing signs of hope for recovery from decades of economic and environmental failure as a result of the recent development of small, local enterprises working with people who live in the inner cities and who have previously suffered from massive unemployment. He points out that small businesses of 500 or fewer employees, many of which are local businesses, are responsible for more than three-quarters of the country's recent job growth; and very small companies of four or fewer employees created more than a third of all new jobs in the first half of the 1990s.[47] Because of its documented success stories, the going local movement is able to be a forum for some of the most meaningful discussions about the important topics of our time.

Another such forum comes in the form of the discussions, books and magazines, and conferences held by people who work to protect the environment and critique the ways that our economy affects nature. Many

nonprofit environmental organizations take part in these discussions, and
they are joined by the members of the public who purchase their writings,
attend their meetings, and discuss those topics at the dinner table. The size
of their numbers and the strength of their commitment has given environmental topics an important place in our national political agenda, and
has compelled local, state, and national governments to work toward conservation. The meetings and writings of nonprofit environmental organizations constitute one of the largest forums in America in which we can
talk consistently about our changing lives and about what those changes
mean for nature. Maybe the work of those nonprofit organizations even
constitutes the largest such forum that we have.

One of the topics that environmental groups are debating the most
passionately is the conversion of rural areas into sprawling strip malls and
subdivisions. This debate has escalated recently to gain full status as a
political issue both nationally and in many states. It was the subject of ballot initiatives in many places during the 1998 elections. If the antisprawl
movement keeps gaining momentum as fast as it did at the end of the
1990s, then it is likely to become the issue that goes the farthest toward
injecting aspects of our personal lives, like being stuck in traffic or losing
the open spaces that we like to see, into mainstream politics.

A prime example is New Jersey, a place with a national reputation for
sprawl, but where voters recently chose to preserve at least half of the
state's remaining 2 million acres of farmland and forest against oncoming
tides of shopping strips, subdivisions, and office parks. Voters sided over-
whelmingly with their governor's proposal to spend close to $1 billion
over the next decade to finance bond issues to buy endangered lands out-
right, or pay owners to give up development rights. Some 68 percent
voted for the proposal.[48]

Maryland governor Parris Glendening analyzed the political trend in
his state, saying, "In both the Maryland primary and general election, voters expressed dissatisfaction with sprawl by defeating candidates they
thought had done or would do too little about controlling growth. That's
a fundamental shift." The shift did not fall according to partisan lines,
however. Glendening is a Democrat. But in two northern Virginia counties, Prince William and Fauquier, upstart Republican candidates won
their races by calling Democratic opponents soft on growth.[49]

Timothy Egan wrote in the *New York Times* that "voters across the country and across party lines, from desert suburbs in the West to leafy cul-de-sacs in the East, voted to stop the march of new malls, homes and business parks at the borders of their communities, and to tax themselves to buy open space as a hedge against future development."[50] Politics in the past has been about topics like jobs. In politics the phrase has been, "It's the economy, stupid." So what can be the source of this new desire to stop the march of new malls? George Bush campaigned and won on a slogan of "no new taxes." So what can be the source of a new willingness of voters to choose to tax themselves to buy open space? This is an unexpected turn of events.

Coming of Age

The source of these changes is that politics is beginning to evolve. It is beginning to take on new subjects, and these new subjects are the personal changes in how Americans live. Politics is beginning to encompass the repetitiveness of our suburban landscape. Politics is beginning to respond to the frustration of sitting in traffic. Local desires are beginning to crawl out from under seemingly endless, and identical, chain stores. These sprouting new politics can be seen in many places. Nationwide, voters approved nearly 200 state and local ballot initiatives on curbing sprawl in 1998.[51]

Even in southern California, famous for automobile culture and sprawl, voters approved a series of urban boundaries around the fast-growing new cities of Ventura County, wedged between Los Angeles and Santa Barbara, and stripped their elected supervisors of the power to approve new subdivisions. The *New York Times* reports that from now on, developers in Ventura County will have to get voter approval before they can push a flood of tile-roofed subdivisions any further onto land that has some of the last big lemon groves in California. About 80 percent of the county will be off-limits to developers, unless voters say otherwise. The *Los Angeles Times* heralded the vote as a "revolution."[52]

Proponents of sprawl, and there are many, say they are stunned by how quickly suburban growth has become a pejorative. "We seem to be at a point now where the word sprawl has been totally demonized," said Clayton Traylor, vice president for political issues for the National Association of Home Builders, which has 195,000 members.[53]

Their surprise is understandable. The history of the United States is in significant part the history of building, the history of spreading out, the history of freedom to put up what enterprises entrepreneurs choose. To turn against this tide is to open a new chapter in American history. As Neal Peirce wrote in the *Washington Post*, "Especially through the Midwest, the Mid-South, Texas and the Mountain States, the very thought of government control on land use has been anathema." But now voters themselves are passing initiatives to control land use.[54]

Some individuals are beating their governments to the punch on these changes. While the governments of places like Tennessee and New Jersey are mandating changes in their cities and states, many individuals are simply making those changes happen without being told. Many of them are choosing, for example, to move out of the suburbs and back into urban centers. These people do not have to move to urban areas; many of them are selling large homes in the suburbs and paying high prices to live in renovated apartments downtown.

This is part of a national trend, according to a new study of twenty-four large American cities by the Brookings Institution and the Fannie Mae Foundation research groups. It reverses a trend that started just after World War II. By 2010, downtown populations are expected to quadruple in Houston to 9,500 people; to more than triple in Cleveland, to 21,000; and to nearly triple in Denver, to 9,250. Chicago's downtown population is expected to increase by a third, to 152,000.[55]

Moving downtown is not for all people. Those choosing to go back to the cities are almost always people who do not have children living at home, for example. They are "empty nesters," childless couples, older people, and singles. But changes in American demographics are going to make these groups a larger part of the national population in the future, and that is going to increase the pools of people who are drawn to cities. By 2010, 72 percent of American households will not have children at home, according to the Brookings study. The "demographic" of people who are becoming attracted to downtown areas is likely to supply an abundance of people to move there.[56]

Governments are supporting the trend of moving back to the cities.

Downtown Denver spent $1 billion on residential construction to build and renovate 1,334 apartments downtown in 1999. That is more than were built there in the last four years combined. Denver is also putting $1 billion into three sports stadiums, an aquarium, and urban shopping complexes to draw additional people downtown and support those already moving there.[57]

Through initiatives like these, voters and governments are actually trying to undo some of the trends that have led us to travel 24 miles in a day on average, and to spread ourselves out. Many of those people living downtown will be able to find both groceries and entertainment within walking distance. Some of them will find that they do not even need to own a car, the way many people in Manhattan today do not have a car, and the way many people in cities of Europe, Asia, and Africa never have. Most of these initiatives, however, have focused on issues like sprawl; too few have addressed new issues like our tendencies to move often to new jobs and new places, to live alone, and to move fast.

We have reached a very strange time for politics. We have almost banished some age-old problems. Historical horrors such as smallpox are gone. We have gotten basic sanitation to most people (though not all). We have a relatively free press and many other crucial elements of a free country. But now we have new problems and new issues, and we will have to address those.

Currently, we do not even have much of the vocabulary needed for a discussion of some of these new issues, nor of how the country is really doing. We talk about "growth" without saying what the growth consists of, for example. Politicians like President Clinton talk of the "longest peacetime expansion of our history." But, as Jonathan Rowe of Redefining Progress says, we have to know what is in that expansion to know whether we want it. He says, "A lot of things can grow, and do. Waistlines grow. Medical bills grow. Traffic, debt, and stress all grow. We can't know whether an 'expansion' is good or not unless we know what it includes. Yet the President didn't tell, and the media hordes didn't ask."[58]

Much of our national dialogue takes place through undefined terms like "expansion." In the words of Jonathan Rowe and Judith Silverstein, "We won't even get to [the topics we care about] unless we start talking about the economy as it is, rather than the way economists tend to think about it." The next time a politician promises "growth," we have to ask, Exactly what is going to double? Traffic? Jet Skis? "People don't experience 'growth.' They experience the things that growth consists of." These are the kinds of questions that we will have to ask in the future, at least if we want to follow our own choices rather than advertising slogans or arcane economic calculations like the ones in the *Index of Leading Economic Indicators*.[59]

It's the beginning of the millennium, and we travel an average of 24 miles a day. We move to new houses every three years or so, on average. This is a big part of who we are. Many of us have a lot of space to live in, or at least more space than our ancestors had. Many of us have little permanence in our lives, since we move frequently and change jobs often. Many of us do not live where we grew up. We have lost many of our childhood friends, and have moved away from many of the people whose decisions we valued in the past. These rapid changes have exposed us to a lot of new places, new choices, and new information.

Sometimes just adjusting to these new places and choices, and keeping up with the newspaper, the magazines, talk radio, and the latest books is beyond the capacity of anyone who adjusts and reads at the pace of a human being. Consider the escalation in how much information is available to us. In 1472 the library at Queens' College in Cambridge, England, had precisely 199 books. At the height of the Renaissance there were people who could claim with some plausibility to have read every important book ever written, according to author Joel Achenbach. The essayist and statesman Francis Bacon once complained that "the whole stock [of books], numerous as it appears at first view, proves on examination to be but scanty."[60]

But this is not the case today. Achenbach says that "the world of knowledge [has become] a vast ocean, horizonless, plunging to impossible depths. The best you can do is occasionally go for a swim." By his count, more than 50,000 books are published every year in America alone. The number of different journals published globally is estimated at 400,000. And, Achenbach says, the World Wide Web that coalesced in 1992 now has millions of sites.[61]

In order to discuss our world, and the changes to our country, Americans are expected to have digested a remarkable amount of writing, speech, and discussion. The difficulty of assimilating it all and forming it into a coherent analysis of our time—and then of applying that analysis in ways that can improve our politics and our communities—may be so large as to have prevented it from happening. One of the reasons why we do not talk about much of what is happening may be that we cannot assimilate all of what is happening and make sense of it.

Some people say that the world is becoming incomprehensible. James Billington, the librarian of Congress, says that "the complaint of every university professor is that there is no way to keep up to date with the research even in a narrowly defined field. At academic conferences, people with Ph.D.s find themselves baffled by the lectures of their colleagues. The expertise is too extreme. There is no common body of knowledge, no common language." He says, "it's the Tower of Babel syndrome."[62]

It is possible, though, that we can deal with this mass of knowledge. One way to deal with this abundance is to look back to what is right in front of us, to stick to simple subjects and focus on our daily activities, like how many places we went during the day. How many people we talked to. How many homes we have lived in since college. How many jobs we have had. Who we live with. These subjects will always be important to all of us. These are the basics. We may have gotten ahead of ourselves with all of our rapid travels. The politics of daily life can bring us back to what is around us.

Such a relatively simple dialogue, spoken in small words and with plain language, can go a long way toward analyzing the changes in America today. Many of the same issues that were important to the people who founded the United States are important to us today, and will always be important to people everywhere. We can talk about them in terms that everyone experiences every day.

By focusing on those topics of daily life, we may be able to begin to see the effects that the invisible strings of subsidies and taxes have on our lives. We may be able to hold politicians and bureaucracies more accountable. We may begin to form constituencies for new policies that suit the ways that we want to live. Such a focus might change our institutions more than most political movements ever have. A discussion, in plain language, about our daily lives might actually be a powerful strategy for political change. It is not partisan. But few people will be able to rebut the statements made by individuals speaking from their hearts about their own lives. So far, Washington, D.C., has heard many statements in public-policy-speak, and remarkably few in plain English.

Notes

1. Jonathan Franzen, "Imperial Bedroom: The Real Problem with Privacy? We Have Too Much of It," *New Yorker*, October 12, 1998.

2. Cliff Cobb, Redefining Progress, private communication, San Francisco, 1997.

3. Ted Halstead, "Foreword," in Clifford Cobb, Ted Halstead, and Jonathan Rowe, *The Genuine Progress Indicator: Summary of Data and Methodology* (San Francisco: Redefining Progress, September 1995).

4. Robert F. Kennedy in a speech at the University of Kansas, March 18, 1968, quoted from John V. Kellenberg, "Accounting for Natural Resources: Ecuador 1971–1990," Ph.D. diss., Johns Hopkins University, Baltimore, Md., 1995.

5. Clifford Cobb, Ted Halstead, and Jonathan Rowe, "If the Economy Is Up, Why Is America Down?" *Atlantic Monthly*, October 1995.

6. Ibid.

7. Cobb, Halstead, and Rowe, *Genuine Progress Indicator*.

8. Ibid.

9. Kate Besleme, Elisa Maser, and Judith Silverstein, "A Community Indicators Case Study: Addressing the Quality of Life in Two Communities," November 1998, Redefining Progress, San Francisco.

10. Clifford W. Cobb and Craig Rixford, *Lessons Learned from the History of Social Indicators* (San Francisco, Calif.: Redefining Progress, 1998).

11. Ibid.

12. *Proceedings of the Colorado Forum on National and Community Indicators*, November 22–23, 1996 (San Francisco: Redefining Progress, 1997).

13. Patricia Gober, *Americans on the Move*, Population Reference Bureau, Population Bulletin, November, 1993; Patricia Gober, "Americans Losing Their Get-Up-and-Go?" Population Reference Bureau News Release, January 13, 1994.

14. Philip J. Longman, "Who Pays for Sprawl?" *U.S. News & World Report*, April 27, 1998.

15. Ibid.

16. James J. MacKenzie, "A Look At Where Cheap Gas Takes Us," *Washington Post*, March 7, 1999.

17. Longman, op. cit.
18. David Roodman, *The Natural Wealth of Nations* (New York: W.W. Norton, 1998); Norman Myers with Jennifer Kent, *Perverse Subsidies: Tax Dollars Undercutting Our Economies and Environments Alike* (Winnipeg, Manitoba, Canada: International Institute for Sustainable Development, 1998).
19. Myers, op. cit.
20. Ibid.
21. Ibid.
22. Norman Myers, "Has the World Gone Mad?" *Guardian,* June 3, 1998.
23. Norman Myers, "Lifting the Veil on Perverse Subsidies," *Nature,* March 26, 1998.
24. Brian Dunkiel, M. Jeff Hamond, and Jim Motavalli, "Sharing the Wealth: If We Shift the Tax Burden from Work to Waste, Everyone Benefits," *E: The Environmental Magazine,* March–April 1999.
25. Hanno Beck, Brian Dunkiel, and Gawain Kripke, *Citizens' Guide to Environmental Tax Shifting* (Washington, D.C.: Friends of the Earth, 1998).
26. Dunkiel, Hamond, and Motavalli, "Sharing the Wealth."
27. Ibid.
28. Beck, Dunkiel, and Kripke, *Citizen's Guide.*
29. Ibid.
30. Ibid.
31. Dunkiel, Hamond, and Motavalli, "Sharing the Wealth."
32. Beck, Dunkiel, and Kripke, *Citizen's Guide.*
33. Ibid.
34. David Roodman, op. cit.
35. M. Jeff Hamond, Stephen J. DeCanio, Peggy Duxbury, Alan H. Sanstad, and Christopher H. Stinson, *Tax Waste, Not Work: How Changing What We Tax Can Lead to a Stronger Economy and a Cleaner Environment* (San Francisco: Redefining Progress, 1997).
36. Neal R. Peirce, "Do Widened Roads Create Their Own Gridlock?" *Washington Post,* January 24, 1999.
37. Ibid.
38. Ibid.
39. Ibid.
40. Ibid.
41. Ibid.
42. Julia A. Heath, "Social Capital and Economic Well-Being: The State of the

American Family in a Global Labor Environment," unpublished paper, University of Memphis, Tennessee, 1998.

43. Ibid.

44. Maryann Cusimano, ed., *Beyond Sovereignty* (Boulder: St. Martin's Press, 1999).

45. Duane Elgin, *Voluntary Simplicity: Toward a Way of Life That Is Outwardly Simple, Inwardly Rich* (New York: William Morrow and Company, 1981).

46. Michael Shuman, *Going Local: Creating Self-Reliant Communities in a Global Age* (New York: Free Press, 1998).

47. Michael H. Shuman, *Small is Lucrative,* unpublished remarks at the International Hearing for the Presidency of the European Union, December 3–4, 1999.

48. Neal R. Peirce, "Sprawl Control: A Political Issue Comes of Age," *Washington Post,* November 15, 1998.

49. Ibid.

50. Timothy Egan, "The Nation Dreams of Fields; The New Politics of Urban Sprawl," *New York Times,* November 15, 1998.

51. Ibid.

52. Ibid.

53. Egan, op. cit.

54. Neal R. Peirce, "Curbing Sprawl: Tennessee's Surprise Breakthrough," *Washington Post,* October 11, 1998.

55. James Brooke, "Denver Stands Out in Trend toward Living in Downtown," *New York Times,* December 29, 1998.

56. Ibid.

57. Ibid.

58. Jonathan Rowe and Judith Silverstein, "The GDP Myth: Why 'growth' isn't always a good thing," *Washington Monthly,* March 1999.

59. Ibid.

60. Joel Achenbach, "The Too-Much-Information Age: Today's Data Glut Jams Libraries and Lives. But Is Anyone Getting Any Wiser?" *Washington Post,* March 12, 1999.

61. Ibid.

62. Ibid.

Index